REGULATORY OVERSIGHT OF AGEING MANAGEMENT AND LONG TERM OPERATION PROGRAMME OF NUCLEAR POWER PLANTS

The following States are Members of the International Atomic Energy Agency:

AFGHANISTAN
ALBANIA
ALGERIA
ANGOLA
ANTIGUA AND BARBUDA
ARGENTINA
ARMENIA
AUSTRALIA
AUSTRIA
AZERBAIJAN
BAHAMAS
BAHRAIN
BANGLADESH
BARBADOS
BELARUS
BELGIUM
BELIZE
BENIN
BOLIVIA, PLURINATIONAL
  STATE OF
BOSNIA AND HERZEGOVINA
BOTSWANA
BRAZIL
BRUNEI DARUSSALAM
BULGARIA
BURKINA FASO
BURUNDI
CAMBODIA
CAMEROON
CANADA
CENTRAL AFRICAN
  REPUBLIC
CHAD
CHILE
CHINA
COLOMBIA
COMOROS
CONGO
COSTA RICA
CÔTE D'IVOIRE
CROATIA
CUBA
CYPRUS
CZECH REPUBLIC
DEMOCRATIC REPUBLIC
  OF THE CONGO
DENMARK
DJIBOUTI
DOMINICA
DOMINICAN REPUBLIC
ECUADOR
EGYPT
EL SALVADOR
ERITREA
ESTONIA
ESWATINI
ETHIOPIA
FIJI
FINLAND
FRANCE
GABON

GEORGIA
GERMANY
GHANA
GREECE
GRENADA
GUATEMALA
GUYANA
HAITI
HOLY SEE
HONDURAS
HUNGARY
ICELAND
INDIA
INDONESIA
IRAN, ISLAMIC REPUBLIC OF
IRAQ
IRELAND
ISRAEL
ITALY
JAMAICA
JAPAN
JORDAN
KAZAKHSTAN
KENYA
KOREA, REPUBLIC OF
KUWAIT
KYRGYZSTAN
LAO PEOPLE'S DEMOCRATIC
  REPUBLIC
LATVIA
LEBANON
LESOTHO
LIBERIA
LIBYA
LIECHTENSTEIN
LITHUANIA
LUXEMBOURG
MADAGASCAR
MALAWI
MALAYSIA
MALI
MALTA
MARSHALL ISLANDS
MAURITANIA
MAURITIUS
MEXICO
MONACO
MONGOLIA
MONTENEGRO
MOROCCO
MOZAMBIQUE
MYANMAR
NAMIBIA
NEPAL
NETHERLANDS
NEW ZEALAND
NICARAGUA
NIGER
NIGERIA
NORTH MACEDONIA
NORWAY

OMAN
PAKISTAN
PALAU
PANAMA
PAPUA NEW GUINEA
PARAGUAY
PERU
PHILIPPINES
POLAND
PORTUGAL
QATAR
REPUBLIC OF MOLDOVA
ROMANIA
RUSSIAN FEDERATION
RWANDA
SAINT LUCIA
SAINT VINCENT AND
  THE GRENADINES
SAMOA
SAN MARINO
SAUDI ARABIA
SENEGAL
SERBIA
SEYCHELLES
SIERRA LEONE
SINGAPORE
SLOVAKIA
SLOVENIA
SOUTH AFRICA
SPAIN
SRI LANKA
SUDAN
SWEDEN
SWITZERLAND
SYRIAN ARAB REPUBLIC
TAJIKISTAN
THAILAND
TOGO
TRINIDAD AND TOBAGO
TUNISIA
TURKEY
TURKMENISTAN
UGANDA
UKRAINE
UNITED ARAB EMIRATES
UNITED KINGDOM OF
  GREAT BRITAIN AND
  NORTHERN IRELAND
UNITED REPUBLIC
  OF TANZANIA
UNITED STATES OF AMERICA
URUGUAY
UZBEKISTAN
VANUATU
VENEZUELA, BOLIVARIAN
  REPUBLIC OF
VIET NAM
YEMEN
ZAMBIA
ZIMBABWE

SAFETY REPORTS SERIES No. 109

REGULATORY OVERSIGHT OF AGEING MANAGEMENT AND LONG TERM OPERATION PROGRAMME OF NUCLEAR POWER PLANTS

INTERNATIONAL ATOMIC ENERGY AGENCY
VIENNA, 2022

COPYRIGHT NOTICE

All IAEA scientific and technical publications are protected by the terms of the Universal Copyright Convention as adopted in 1952 (Berne) and as revised in 1972 (Paris). The copyright has since been extended by the World Intellectual Property Organization (Geneva) to include electronic and virtual intellectual property. Permission to use whole or parts of texts contained in IAEA publications in printed or electronic form must be obtained and is usually subject to royalty agreements. Proposals for non-commercial reproductions and translations are welcomed and considered on a case-by-case basis. Enquiries should be addressed to the IAEA Publishing Section at:

Marketing and Sales Unit, Publishing Section
International Atomic Energy Agency
Vienna International Centre
PO Box 100
1400 Vienna, Austria
fax: +43 1 26007 22529
tel.: +43 1 2600 22417
email: sales.publications@iaea.org
www.iaea.org/publications

© IAEA, 2022

Printed by the IAEA in Austria
June 2022
STI/PUB/1973

IAEA Library Cataloguing in Publication Data

Names: International Atomic Energy Agency.
Title: Regulatory oversight of ageing management and long term operation programme of nuclear power plants / International Atomic Energy Agency.
Description: Vienna : International Atomic Energy Agency, 2022. | Series: IAEA safety reports series, ISSN 1020–6450 ; no. 109 | Includes bibliographical references.
Identifiers: IAEAL 22-01493 | ISBN 978–92–0–108122–3 (paperback : alk. paper) | ISBN 978–92–0–108222–0 (pdf) | ISBN 978–92–0–108322–7 (epub)
Subjects: LCSH: Nuclear power plants — Maintainability. | Nuclear power plants — Safety measures. | Nuclear power plants — Reliability.
Classification: UDC 621.039.58 | STI/PUB/1973

FOREWORD

At present, a large number of nuclear power plants are completing their originally licensed lifetime and are in the process of extending their operational life by taking into account the assessment of the residual life of the vital components, equipment replacements or refurbishments. The safety of nuclear power plants during long term operation (LTO) has become more important owing to the increase in the number of licensees giving high priority to assessments for continuing operation of nuclear power plants beyond the time frame originally anticipated in the design. Accordingly, the regulatory framework for oversight of LTO has to take into account relevant safety aspects as an important basis for safe LTO.

The process for how LTO is to be implemented depends on the regulatory background and approaches applied in the Member State. However, the regulatory oversight and tasks related to ageing management and preparedness for, and implementation of, the LTO programme starts from the overall plans, and lasts through the preparatory activities and formal authorization of LTO to the period when nuclear power plants commence operation in the extended period. This Safety Report provides practical information based on existing regulatory practices of the Member States for regulatory oversight of ageing management, including equipment qualification, design modifications, replacement or refurbishment of structures, systems and components for ensuring safe LTO of nuclear power plants. This Safety Report addresses the regulatory framework, including regulations, regulatory requirements and guides, regulatory processes, practices applied for the oversight, competence and preparation of the regulatory body for oversight of the plant preparedness for, and implementation of, an LTO programme and other plant programmes with respect to ageing management. This Safety Report provides information for nuclear safety authorities, operating organizations, licensees, manufactures, designers and technical support organizations of the IAEA Member States considering authorization for LTO of currently operating nuclear power plants.

This Safety Report complements the IAEA Safety Standards Series No. GSR Part 1 (Rev. 1), Governmental, Legal and Regulatory Framework for Safety; No. SSR-2/2 (Rev. 1), Safety of Nuclear Power Plants: Commissioning and Operation; No. SSG-48, Ageing Management and Development of a Programme for Long Term Operation of Nuclear Power Plants; and No. GSG-13, Functions and Processes of the Regulatory Body for Safety.

The contributions of all those who were involved in the drafting and review of this publication are greatly appreciated. The IAEA officer responsible for this publication was G. Petofi of the Division of Nuclear Installation Safety.

EDITORIAL NOTE

CONTENTS

1. INTRODUCTION

1.1. BACKGROUND

IAEA Safety Standards Series No. GSR Part 1 (Rev. 1), Governmental, Legal and Regulatory Framework for Safety [1], states in para. 4.40 that:

"[t]he regulatory body shall review and assess the particular facility or activity in accordance with the stage in the regulatory process (initial review, subsequent reviews, reviews of changes to safety related aspects of the facility or activity, reviews of operating experience, or reviews of long term operation, life extension, decommissioning or release from regulatory control). The depth and scope of the review and assessment of the facility or activity by the regulatory body shall be commensurate with the radiation risks associated with the facility or activity, in accordance with a graded approach."

According to GSR Part 1 (Rev. 1) [1], it is the task of the regulatory body to provide for safety oversight of facilities throughout the operational lifetime of the facilities. Requirement 32 states: "**[t]he regulatory body shall establish or adopt regulations and guides to specify the principles, requirements and associated criteria for safety upon which its regulatory judgements, decisions and actions are based**" [1]. Requirement 25 stipulates that "**review and assessment of information shall be performed prior to authorization and again over the lifetime of the facility or the duration of the activity**" to determine whether the requirements and conditions for the authorization have been met.

IAEA Safety Standards Series No. SSR-2/1 (Rev. 1), Safety of Nuclear Power Plants: Design [2], and No. SSR-2/2 (Rev. 1), Safety of Nuclear Power Plants: Commissioning and Operation [3], require that the plant design take due account of ageing of the structures, systems and components (SSCs), and that an effective ageing management programme (AMP) be implemented during operation, to ensure that the safety functions of SSCs are fulfilled over the entire operating lifetime of the plant. In relation to long term operation (LTO), SSR-2/2 (Rev. 1) [3] requires that, where applicable, a comprehensive programme for LTO be established and implemented on the basis of safety assessments, with due consideration of the ageing of SSCs.

IAEA Safety Standards Series No. SSG-48, Ageing Management and Development of a Programme for Long Term Operation of Nuclear Power Plants [4], provides recommendations for meeting the related Requirements 14 and 16 of SSR-2/2 (Rev. 1) [3]. It outlines the major arrangements, steps and items that the operating organization needs to implement in order to provide a sound AMP and prepare for the safe LTO of the plant. Paragraph 7.39 SSG-48 [4] states: "the regulatory body should oversee, that the safety of the nuclear power plant will be maintained throughout the period of long term operation in accordance with current safety standards and national regulatory requirements."

IAEA Safety Standards Series No. GSG-13, Functions and Processes of the Regulatory Body for Safety [5], in its Appendix III (paras III.1–III.15), provides recommendations on subjects of review and assessment by the regulatory body on equipment qualification and management of ageing. IAEA Safety Standards Series No. GSG-6, Communication and Consultation with Interested Parties by the Regulatory Body [6], contains general recommendations on communication and consultation with interested parties by the regulatory body for all facilities and activities, for all stages in their lifetime.

The methodology described in IAEA Safety Standards Series No. SSG-25, Periodic Safety Review for Nuclear Power Plants [7], is used by many Member States, among other tools, to assess the cumulative effects of plant ageing and plant modifications. It provides recommendations on some aspects of ageing of SSCs, mainly related to non-physical ageing, and can also be used to seek out and implement safety improvements if the plant is to be operated beyond its original design life. In SSG-25 [7], the safety factors 2: 'Actual condition of SSCs important to safety'; 3: 'Equipment qualification'; and 4: 'Ageing', focus attention on the aspects of ageing of plant SSCs. The guide also addresses briefly the roles and responsibilities of the regulatory body in conjunction with the periodic safety review (PSR); however, the detailed aspects of the regulatory oversight process for LTO are not explained.

To complement the above requirements and recommendations, in 2009 the IAEA initiated the Extrabudgetary Programme on International Generic Ageing Lessons Learned (IGALL) with the objective of developing a general framework to effectively capture and disseminate experience and lessons learned in relation to ageing management. During the first phase of the IGALL programme (2010–2013), the participating Member States systematically summarized the ageing related research results and operating experience in Safety Reports Series No. 82, Ageing Management for Nuclear Power Plants: International Generic Ageing Lessons Learned (IGALL) [8]. Reference [8] contains a generic sample of ageing management review (AMR) tables, a collection of proven AMPs and a collection of typical time limited ageing analyses (TLAAs) for in-scope SSCs. However, Ref. [8] does not address the relevant regulatory aspects for the oversight of ageing management.

In Phase 3 of the IGALL Programme, conducted in 2016–2017, the participating Member States recognized that, in addition to gathering the operating experience and practice of the operating organizations, the regulatory framework applicable for the safe LTO of nuclear power plants is also an important area where relevant national regulatory approaches and practices can be assimilated and shared.

1.2. OBJECTIVE

The objective of this Safety Report is to provide technical and practical information based on the existing regulatory approaches and practices of Member States concerning safety oversight of ageing management and LTO of nuclear power plants. It is also meant to support the application of relevant IAEA safety requirements, such as GSR Part 1 (Rev. 1) [1], in terms of regulatory oversight of the safety requirements included in SSR-2/1 (Rev. 1) [2] and SSR-2/2 (Rev. 1) [3], and the recommendations included in SSG-48 [4] and SSG-25 [7].

The information contained in this Safety Report is based on the Member States' practices in the regulatory oversight of the plants' AMPs and the preparedness for, and implementation of, LTO. The Safety Report describes the practices followed by the Member States in defining and/or applying:

(a) Regulatory requirements, guidance for ageing management and LTO;
(b) Preconditions of the regulatory body for LTO, including documentation to be submitted, scheduling of submissions and applicable review and assessment processes;
(c) Various authorization processes applied to LTO;
(d) Regulatory oversight of ageing management and other plant programmes with respect to LTO;
(e) Regulatory practices applied for the oversight of plant preparedness for and implementation of LTO;
(f) Documentation during regulatory oversight of LTO.

1.3. SCOPE

This Safety Report compiles the regulatory practices in Member States operating nuclear power plants for oversight of ageing management and preparedness for safe LTO. The information provided in this Safety Report is primarily relevant for regulatory bodies that perform safety oversight of operating plants considering authorization for LTO or plants already in LTO. The information is also useful for operating organizations, licensees, manufacturers, designers and technical support organizations (TSOs).

Preparation for ageing management starts in the design phase to make adequate provision to facilitate effective ageing management during plant fabrication, transportation, construction, commissioning, operation (including LTO) and decommissioning. The regulatory bodies need to be prepared to perform the safety oversight of ageing management throughout all stages of the lifetime. This publication specifically discusses regulatory issues of ageing management from the perspective of LTO.

Managing ageing for nuclear power plants involves ensuring the availability of the safety functions throughout the service life of the plant, taking into account the changes in the operating and environmental conditions and/or properties of relevant SSCs that occur over time. This will involve addressing both the effects of physical ageing of SSCs, resulting in degradation of their performance characteristics, and non-physical ageing (obsolescence) of SSCs, that is their becoming out of date with respect to current knowledge, codes, standards, technology and applicable regulatory requirements. This Safety Report discusses regulatory practices to cover physical ageing and to address the technological obsolescence of SSCs proactively within the scope of ageing management. In line with para. 2.29 of SSG-48 [4], "conceptual aspects of obsolescence, such as obsolescence of knowledge and compliance with current regulations, codes and standards", are not addressed in this publication.

This Safety Report does not cover the regulatory framework applied by Member States for the overall regulatory oversight of nuclear power plants comprehensively, but some general elements of regulatory oversight, including applicable regulatory documents, competence of regulatory staff, the management system, inspection and review and assessment practices, are addressed as an integral approach of regulatory oversight concerning ageing management and LTO (see Requirements 18, 19 and 25–29 of GSR Part 1 (Rev. 1) [1]).

This Safety Report can also be applied as a reference for oversight of ageing management and preparedness for LTO of other nuclear installations[1] using a graded approach, with due consideration of the differences in hazard potential and complexity of affected systems.

---

[1] According to the IAEA Safety Glossary (2018 Edition) [9]:
— The definition of the term 'nuclear installation' includes: nuclear power plants; research reactors (including subcritical and critical assemblies) and any adjoining radioisotope production facilities; storage facilities for spent fuel; facilities for the enrichment of uranium; nuclear fuel fabrication facilities; conversion facilities; facilities for the reprocessing of spent fuel; facilities for the predisposal management of radioactive waste arising from nuclear fuel cycle facilities; and nuclear fuel cycle related research and development facilities.
— The term 'nuclear facility' is defined as a facility in which nuclear material is produced, processed, used, handled, stored or disposed of.

The information provided in this Safety Report represents the proven practices of the Member States that participated in the development of this publication. However, other Member States for which all aspects might not be directly applicable can draw lessons from this Safety Report for their own use, as appropriate.

In this Safety Report, it is assumed that the requirements and recommendations described in Section 1.1 are known and implemented in the Member States, and so they are not repeated unless additional information is provided or variances in the application need to be discussed.

1.4. STRUCTURE

This Safety Report is divided into 10 sections:

— Section 1 (Introduction) provides the background, objective, scope and structure of this report.
— Section 2 (Legal framework for safety in Member States) describes the levels of legal systems of Member States where ageing management and LTO can be discussed.
— Section 3 (Regulatory requirements considered as preconditions for LTO) contains guidance on providing requirements and conditions to be met by the licensee before applying for an LTO licence.
— Section 4 (National requirements and guidance for the LTO authorization process) provides guidance on regulating the authorization process, including review and assessment to be followed for LTO licensing.
— Section 5 (National requirements and guidance for plant documentation and programmes relevant for LTO) explains the expectations on documentation, how to deal with management of modifications and configuration management with respect to LTO, including the final safety analysis report (FSAR) and other document updates, and the expectations concerning the plant programmes that need to be appropriately aligned with the regulatory requirements for ageing management and LTO.
— Section 6 (National requirements and guidance for ageing management for LTO) provides information for expectations on scope setting, AMPs, AMRs, revalidation of TLAAs and technological obsolescence.
— Section 7 (Periodic safety review with respect to LTO) describes the role and conduct of PSRs for LTO, where appropriate.
— Section 8 (Preparation of the regulatory body for LTO programme review) describes the practices applied for the self-preparation of the regulatory body for oversight of LTO.

— Section 9 (Oversight of LTO programme preparation and implementation) contains guidance for regulators on reviewing the scope setting, AMR, AMP development and implementation TLAA revalidation, and on documentation and follow-up of findings.
— Section 10 (Specific activities of the regulatory body during implementation of LTO) describes the regulatory practices followed in the LTO period.

2. LEGAL FRAMEWORK FOR SAFETY IN MEMBER STATES

This section describes the considerations the legal framework of a Member State needs to address to facilitate effective preparation for, and implementation of, LTO by the plant licensee(s).

As specified in Requirement 2 of GSR Part 1 (Rev. 1) [1], "**[t]he government shall establish and maintain an appropriate governmental, legal and regulatory framework for safety within which responsibilities are clearly allocated**", to ensure safe operation of nuclear facilities and conduct of associated activities. The legal authority for the statutory obligation of regulatory control over nuclear facilities and activities of the licensees is conferred on the regulatory body established through the legal system of the Member State.

A basic condition for a regulatory body to perform regulatory oversight of both ageing management and LTO in a nuclear power plant is to have an adequate system of laws and regulatory requirements.

The legal and regulatory framework of most Member States has a pyramid-like hierarchical structure, with the requirements becoming more detailed and/or specific at each level from the top of the pyramid. At the top is the national or federal law or act on safe use of nuclear energy, below which follows a set of other, more detailed, legal instruments, such as government decrees, ordinances, regulations, licences and regulatory documents that have mandatory power and are legally binding.

The legal powers of the regulatory body in different Member States can vary in terms of the level of the legal instrument that establishes it and that specifies the status and authorizations within the state organization or governmental hierarchy. Within that, in most Member States the regulations outline the responsibilities of the government and the regulatory body for authorization and oversight of operation of nuclear power plants, including LTO.

In addition, the regulatory body in many Member States is authorized to issue mandatory decrees and resolutions on licence conditions, which may also be part of the legally binding set of instruments.

As for implementation, the regulatory body in most cases issues other documents to explain or interpret these legally binding requirements in the form of guidance documents, regulatory standards, administrative letters, review plans, official notifications, statements or opinions. Except for some Member States, these are formally non-binding documents, but deviation from their contents typically entails a more rigorous regulatory vigilance and justification in some actions, a series of enforcement steps and/or closer oversight of certain activities to confirm that the operating organization satisfies the legal requirements.

In most Member States, the option of potential LTO had not been foreseen at the time of the original authorization process for the operation of the plant, and so the applicability of the legal framework to cover LTO had to be reviewed by the regulatory body. The most important finding from reviewing the LTO authorization approaches in Member States is that practices differ concerning the question as to whether there is a fixed licensing period option.

Some Member States, such as Hungary, the Republic of Korea[2] and the United States of America, issued operating licences with specified licensing periods, which may be coincident with the original design life, most often 30 or 40 years. Other Member States, such as Canada, issued operating licences for shorter licensing periods (5 to 10 years), and the licensee is obliged to apply for licence renewal at the end of each term. Operating licences in some other Member States are indeterminate, without a specific timeline or expiration date, allowing the licensee to operate the plants as long as they are within the licensing basis. These Member States apply the results of PSR as a justification for allowing operation beyond the original design life or original licensing period. Mixed approaches using some aspects of both methods also exist in some Member States.

Review of the existing legal and regulatory framework within a Member State is important to confirm whether the existing legal instruments and regulatory framework allow for, or are compatible with, LTO. Furthermore, such review ensures that the legal and regulatory framework provides clear goals and that the optimum safety level is required for those nuclear power plants that intend to operate beyond the original design life or original licensing period. If the legal and regulatory framework is not applicable to or is incompatible with LTO, the regulatory body can initiate amendments or make separate regulatory arrangements with the licensee to define the necessary requirements.

Some Member States carried out comprehensive reviews of their current regulatory framework and authorization practices, and on the basis of these necessary amendments were proposed and implemented. This revision process

---

[2] Hungary and the Republic of Korea also require PSR, but LTO licensing is carried out separately from the PSR process.

included all of the necessary levels of the legal and regulatory framework in the Member States.

For other Member States, the regulatory approach focused on assessing SSCs whose ageing could potentially limit the safe operating lifetime of the plant. This formed the basis on which the LTO authorization process was established.

Following the requirements specified in GSR Part 1 (Rev. 1) [1], Member States implement the regulatory requirements for LTO with the intention of maintaining the operation of the plant at a similar or higher safety level for the period of LTO. As described in GSR Part 1 (Rev. 1) [1], Member States implement a system for informing and consulting interested parties during the process of amending regulations. According to the experience of Member States, for PSR for LTO (i.e. extended PSR to justify LTO) or licence renewal applications, it is beneficial to include a preparatory phase to specify the details of the application or documentation (see GSG-13 [5]). Reference [10] describes important considerations and approaches for the authorization process for LTO. The best practice calls for the regulatory requirements for LTO and ageing management to be embedded in all levels of the legal framework (legal basis and regulatory guidance), irrespective of the approach followed by the Member State for the authorization process for LTO.

3. REGULATORY REQUIREMENTS CONSIDERED AS PRECONDITIONS FOR LONG TERM OPERATION

This section discusses the preconditions for LTO that are assumed to have been met before a nuclear power plant enters its LTO period.

Requirements for LTO are typically set at a legal and/or regulatory requirement level, depending on the regulatory system of the Member State. The regulatory bodies can be authorized by a law governing the rules of safe use of nuclear energy, or its supporting decrees or ordinances, to issue detailed requirements and regulatory guides to define the preconditions, preparation and implementation of safe LTO of nuclear power plants.

3.1. REQUIREMENTS OF THE REGULATORY BODY

The legally binding requirements address ageing management, authorization preconditions and the authorization process itself. Requirements 14 and 16 in SSR-2/2 (Rev. 1) [3] and Section 4 of SSG-48 [4] contain the internationally accepted requirements and recommendations for both ageing management and LTO preconditions.

In line with this, the establishment and implementation of an effective AMP is typically required in order to ensure that the safety functions of the SSCs are fulfilled over the entire operating life of the nuclear power plant, including LTO. The following are typically expected from an effective AMP of the plant (see SSG-48 [4]):

(a) Degradation mechanisms and ageing effects are understood for each in-scope SSC, parameters influencing the degradation mechanisms leading to ageing effects are evaluated and assessed using a graded approach.
(b) Proactive behaviour to anticipate all potential ageing related technical issues is applied for ageing management.
(c) An effective AMP has the nine generic attributes enumerated in detail in table 2 of SSG-48 [4].
(d) AMPs are systematically developed, implemented, evaluated and improved.
(e) All ageing management activities, including, for example, the plant programmes and PSR activities, in the plant are coordinated, and effective cooperation of the plant divisions responsible for the individual plant programmes and technical areas is ensured.
(f) Appropriately qualified experts are involved in any aspect of ageing management, including assuming the role of an 'intelligent customer' for services purchased from external TSOs.

In line with the national regulations, the operating organization establishes and implements a comprehensive programme (see para. 4.54 of SSR-2/2 (Rev. 1) [3]) for preparing for LTO and carrying out the necessary assessments and activities for safe LTO.

Well before the start of the LTO authorization process, it is important for the regulatory body to establish and share an overall strategy for the regulatory oversight of LTO to address:

(a) Safety assessments to justify safe LTO with due consideration of ageing;
(b) The role of PSR for justification, where appropriate;
(c) The regulatory review and assessment, approval and authorization process for LTO with requirements on scheduling and contents of documents to be submitted for LTO justification;
(d) The process for safety improvements (Principle 2 of Ref. [11]) (see further details in Section 3.2);
(e) How it will engage with interested parties.

The following basic requirements are typically incorporated into the regulatory framework for the completion of the programme for safe LTO (see SSR-2/2 (Rev. 1) [3], SSG-48 [4] and SSG-25 [7]):

(a) A clear plant policy (see IAEA Safety Standards Series No. GS-G-3.5, The Management System for Nuclear Installations [12]), describing the principles and concepts for LTO and ageing management, exists. The plant personnel are familiar with and properly understand the policy.
(b) Preconditions are met, including an adequate or updated current licensing basis (CLB), safety upgrading and verification, and operational (plant) programmes.
(c) Scopes for ageing management and LTO are set.
(d) Identification and reviews of potential degradation mechanisms and ageing effects are carried out.
(e) Identification and revalidation of TLAAs is done.
(f) A review of AMPs and other plant programmes is completed, and the programmes are enhanced or new programmes are developed, as necessary.
(g) A programme for safe LTO is implemented.
(h) A programme for ensuring human resources and knowledge management for LTO is developed.

The most important safety related issues for LTO to be regulated are ageing management of SSCs together with the relevant plant programmes (see Section 5.6) and ageing related safety analyses. These areas are regulated at the appropriate level of the legal system with the detailed requirements established in lower level mandatory and non-mandatory regulations, guides and standards. More detailed information is provided in Section 6.

Regulatory body requirements can specify the preconditions and other licensing conditions for LTO in more detail to further specify the provisions for PSR, ageing management and other plant programmes, revalidation of TLAAs and AMR. In this way, the regulatory body requirements supplement the higher level regulations with specific, detailed provisions, usually containing technical level approaches.

The requirements in the Member States are to be clarified during the preparation phase of LTO in order for the operating organization(s) to address and fulfil them in a timely way in the LTO programme. Depending on the regulatory system, requirements for ageing management and for other plant programmes are amended and usually made more detailed to allow for the implementation of LTO. More detailed information is provided in Sections 5 and 6.

Some regulatory bodies also require the review and/or setting up of an organizational arrangement in the plants for LTO preparation and implementation and/or an authorized organizational entity assigned by the plant to have responsibilities for ageing management and/or the LTO programme. More detailed information can be found in Section 6.

Section 4 of SSG-48 [4] describes relevant plant documentation and programmes that should be in place for evaluation of the LTO. Some regulatory bodies provide a mandatory list of updated documents or analyses to be submitted for LTO. These documents include but are not limited to:

(a) The latest PSR report with an action plan to improve plant safety;
(b) An updated FSAR;
(c) Updated deterministic and probabilistic safety analyses considering design modifications, refurbishment and replacements;
(d) Updated safety classification lists;
(e) A reassessment of the design bases, with special attention paid to the most safety significant components, such as the reactor pressure vessel and the other main circulation loop components, as appropriate;
(f) AMPs revised for safe LTO during an AMR;
(g) Other plant programmes, including maintenance, corrective action, testing, surveillance, in-service inspection (ISI), equipment qualification and water chemistry programmes revised for LTO during an AMR;
(h) Severe accident management guidelines;
(i) An emergency preparedness and response plan;
(j) A radiation protection programme;
(k) An environmental monitoring programme;
(l) An integrated management system manual;
(m) A fire protection programme;
(n) An organizational structure for LTO;
(o) A radioactive waste management programme;
(p) An initial decommissioning plan;
(q) A report on actions taken against international operating experience feedback (e.g. the IAEA report on the Fukushima Daiichi nuclear power plant accident).

The documents and analyses are typically required to be updated to address LTO; justification for the continued validity of existing documents and analyses is provided for regulatory review. Depending on the type of LTO approval process (licence renewal or PSR for LTO), the regulatory body might require different documents and analyses.

3.2. SAFETY IMPROVEMENTS

In some Member States, the operating organizations of nuclear power plants are required to perform a PSR at least every 10 years. This is an opportunity not only to review the conformity of the plant with its CLB, but also to identify possible safety improvements that would bring the facility up to modern standards. Safety improvements can be based on state of the art technology improvements, the results of research and development (R&D) activities, and operating experiences from the plant's own operation or from other nuclear power plants (see SSG-25 [7]). Examples of using operating experiences for safety improvements include the major accidents of the nuclear industry. When LTO is being considered, safety improvements become even more relevant, since large investments could be more feasible for a longer operational lifetime of the plant. Safety improvements can be related to the plant design, but also to organizational issues (e.g. management systems and procedures and plant programmes). Based on the results of the PSR (or similar assessments), where applicable according to national regulations, the regulatory body decides on the acceptability of continued operation of the plant for the LTO period. More detailed information about PSR is provided in Section 7.

In the case of Member States where a formal licence renewal process exists, safety improvements or upgrades may be a precondition for licence renewal. In Member States that issue unlimited term licences, safety improvements or upgrades might be required to be implemented either before the LTO period starts or within a specified time frame after entering LTO as a licence condition.

The need for urgent safety improvements or upgrades can also occur any time significant issues arise that may put the safety of the plant at risk and therefore need to be addressed without delay. Operating experience feedback, particularly related to incidents or events, can be an important trigger for plant improvements. The safety assessments (or 'stress tests') performed in many Member States after the Fukushima Daiichi accident are a good example of actions performed outside the frame of PSRs. Some Member States reported that the implementation of measures based on the lessons from that accident had been set as a precondition for authorizing LTO.

4. NATIONAL REQUIREMENTS AND GUIDANCE FOR THE LONG TERM OPERATION AUTHORIZATION PROCESS

This section is meant to describe the national requirements and guidance to be in place in the Member States for authorization of LTO depending on the authorization approach applied.

4.1. APPROACHES FOR THE AUTHORIZATION OF LONG TERM OPERATION

The technical steps required by the regulatory body to justify LTO for a nuclear power plant might be part of a licence renewal application and/or reflect specific requirements of the last PSR cycle before the expiration of the assumed original design life combined with an LTO safety review. For the last PSR term prior to entering LTO, an intensified safety review methodology may be developed to focus on ageing management and lifetime assessments (e.g. TLAAs) of in-scope SSCs as prerequisites to obtain an operating licence extension beyond an established time frame.

Although each Member State may have its own LTO justification methods, the various approaches can be grouped into two main categories (see Ref. [10]):

(1) The PSR method is typically used in Member States with unlimited or continuing operating licences (e.g. the Czech Republic, France, Sweden and Switzerland). The requirements specified for PSR and LTO safety review are set for continued operation. The operating licence remains valid as long as safety requirements are met. In some of these Member States, the PSR is used as a tool for regularly reviewing the CLB and identifying safety improvements. Typical features of this approach include:
    (i) Operation including LTO is authorized based on periodic reviews of safety performed by the operating organization and assessed by the regulatory body.
    (ii) The licensing basis is thoroughly reviewed during the PSR process and significant efforts are made to identify safety improvements based on state of the art international standards, good practices, R&D results and operating experience. Although operating experience is continuously collected and assessed, it is systematically trended and reviewed during PSRs.

(iii) Based on the results submitted by the operating organization, the regulatory body can authorize continued operation for the next PSR cycle, which is typically 10 years. The regulatory body reviews, approves and, if necessary, extends the safety improvement actions proposed by the operating organization. Safety improvements can address the design as well as operational and organizational aspects.

(2) Another approach is based on a limited term licence and a licence renewal process (e.g. Canada, Hungary, Romania, Spain, the USA). For example, the Code of Federal Regulation in the USA sets the rules for the renewal process for limited term licences. It takes credit for the plant's compliance with its CLB and adds an additional requirement that the plant demonstrate that the effects of ageing will be managed adequately to ensure continued safe plant operation, that TLAAs have been appropriately revalidated and that an FSAR update to describe the ageing management activities has been performed. The regulator may require safety improvements for LTO with due basis. Among the Member States using a similar process, there are differences regarding the number of subsequent licence renewals and the duration of the extended licence. In some of these Member States, the PSR is also used as a tool for regularly reviewing CLB and identifying safety improvements. Typical features of the licence renewal process include the following:

(i) A strong CLB approach is followed to review and ensure continuous compliance with requirements.

(ii) A formal licence renewal process is applied in the frame of an extended safety review for a designated scope of SSCs. This is a single authorization step based on a licence renewal application to demonstrate that safety requirements will be met during the extended lifetime by adequate management of the effects of ageing on the intended functions of in-scope SSCs. Actions necessary for maintaining the compliance are incorporated into the application.

(iii) The regulatory body reviews the licence renewal application, conducts the appropriate inspections and issues a decision at the end of the process.

4.2. AUTHORIZATION PROCESS

Requirement 24 of GSR Part 1 (Rev. 1) [1] states: "**[t]he applicant shall be required to submit an adequate demonstration of safety in support of an**

application for the authorization of a facility or an activity." For authorization of LTO, the regulatory body performs the following activities:

(a) Issues guidance for the format and content of documents to be submitted;
(b) Reviews and assesses the safety assessment submitted in support of the application using a graded approach;
(c) Verifies the competence of individuals with responsibilities for safety;
(d) Imposes limits, conditions and controls, as appropriate;
(e) Extends, amends, renews, suspends or revokes the former authorization, as appropriate;
(f) Conducts its own assessment, review and inspection, and takes into account operating experience to be able to make a decision on the submission;
(g) Documents the basis for the decision and informs the interested parties about the decision.

Further specific recommendations for the authorization process can be found in GSG-13 [5]. Authorization is generally granted or denied in accordance with the governmental, legal and regulatory framework and covers all stages of the lifetime of the nuclear power plant. Authorization of LTO incorporates all relevant safety aspects of the LTO period.

Depending on the type of licensing in a Member State, there are different procedures for authorization of LTO. In Member States where unlimited operating licences are granted, authorization consists of the review, assessment and then approval of the documentation submitted for LTO (e.g. in the form of the PSR report). In Member States that have implemented a formal licence renewal process, the licensee is typically required to submit a licence renewal application (see e.g. Ref. [13]). In both cases, the authorization for LTO is issued by the regulatory body or by the government based on the submitted safety evidence.

The regulatory body has the responsibility to review the submitted LTO documents. In many Member States, independent TSOs or independent experts are involved in the review. In these instances, the conditions for this third party participation in the process are laid down in the regulations and/or in regulatory guides. The respective IAEA recommendations can be found in IAEA Safety Standards Series No. GSG-12, Organization, Management and Staffing of the Regulatory Body for Safety [14].

The time frame for the authorization process, including review of LTO documents or licence applications at the various stages, varies between Member States. In some Member States, a public hearing process will be required before or during the authorization process. In other Member States, quasi-judicial hearings may be held if suitable issues or concerns are identified by the public.

The rules, process and conditions of such hearings are clearly defined in the legal and regulatory documents.

Principles for LTO authorization are established in the legal and regulatory framework of the Member States. Examples of the main principles of Member States for LTO authorization include the following:

(a) The regulatory framework has to provide clear and detailed guidance on the process and on what needs to be submitted during the LTO authorization process. Authorization of LTO needs to be based on a predefined list of documents. The legal framework for LTO is described in Section 2. The documentation typically required by the regulatory body for approval of LTO is given in Section 3.1.

(b) The regulatory body needs to develop and establish a clear and explicit set of requirements, criteria and standards forming the basis for LTO authorization.

(c) LTO is authorized only when the regulatory body has confirmed, by review and assessment of the submitted documentation, that continued operation of the nuclear power plant does not pose an unacceptable radiation risk to people or the environment. It has to be demonstrated by the licensee that the effects of ageing will be managed adequately and that TLAAs have been appropriately revalidated. This may include confirmation that the applicant has the organizational capability for LTO, the organizational structure, adequate resources, adequate competence of managers and staff, and appropriate management system arrangements to comply with the safety requirements.

(d) The authorization process needs to be transparent to the public, and the associated decisions need to be published or made available to the public by defined means and methods.

In some Member States, there are interactions between the regulatory body and the licensee during the preparatory phase in the form of consultation. In addition, regulatory approval of the LTO preparation programme and an inspection programme during its implementation are also applied. In this case, the first authorizations of different steps of the process may start several years before the LTO period. In other Member States, the formal authorization consists of a one step process, which covers a period of approximately one year.

In some Member States, other state or local authorities also take part in the authorization process in specific areas. For example, authorities responsible for environmental protection, public health, labour safety and industrial safety may be involved in the process. Accordingly, the participation of these authorities may add to the complexity and regulatory assessment period of LTO applications before authorization.

In some Member States, a new or updated environmental impact assessment for LTO might be required in environmental legislation. Environmental authorization of LTO typically precedes the nuclear safety authorization. The regulatory body may be involved in the environmental authorization process. In this case, the regulatory body is usually tasked with reviewing the scenarios determined for analyses of the worst case environmental impact during LTO.

Some Member States have reported that even though the regulatory body and its TSO(s) review the submissions, the authorization itself is granted by other government organizations, such as ministries responsible for nuclear energy or nuclear safety.

The content of the authorization itself differs according to the regulatory framework of the Member States:

(a) If LTO is formally authorized in the framework of the PSR, it is made in the framework of a current operating licence for the next PSR cycle.
(b) If the authorization is a licence renewal or a similar licence, a new operating licence for a specified time period is issued to cover the LTO term. Irrespective of the licence term, PSRs might still be required to be conducted during LTO.

In both cases, the authorization may contain a set of general and specific conditions that reflect the conclusions of the LTO review. Such factors may include the following (see GSG-13 [5] and IAEA Safety Standards Series No. SSG-12, Licensing Process for Nuclear Installations [15]):

(a) A sufficiently detailed description of the nuclear installation;
(b) Operational limits and conditions;
(c) Restrictions from other authorities;
(d) Requirements for notifying the regulatory body of safety related events, a list of reportable events;
(e) Requirements for ensuring the fitness for service of SSCs, including relevant plant programmes (see Sections 5.6 and 6);
(f) Requirements for routine reports on safety of operation;
(g) Requirements for arrangements for emergency preparedness;
(h) Severe accident related conditions;
(i) Requirements for quality assurance and retention of records;
(j) Period or validity of authorization;
(k) Required safety reviews;
(l) Requirements for specific procedures and modes of operation;
(m) Safety related aspects to enable effective regulatory control, including regulatory inspection and enforcement;

(n) ISI requirements;

(o) Fresh fuel management and inventory requirements;

(p) Spent fuel and radioactive waste management requirements;

(q) Modifications to the facility.

5. NATIONAL REQUIREMENTS AND GUIDANCE FOR PLANT DOCUMENTATION AND PROGRAMMES RELEVANT FOR LONG TERM OPERATION

This section addresses the regulatory requirements and guidance for the most crucial plant documentation and programmes that form the basis for updating the FSAR and demonstrating readiness for LTO.

5.1. UPDATING THE FINAL SAFETY ANALYSIS REPORT

In most Member States, an updated version of the FSAR is part of the documents justifying LTO. Member States have requirements related to updating the FSAR, either as a separate document, as part of other documentation or as part of the licence conditions.

In practically all Member States, the FSAR is updated, taking into account the changes in the design basis, renewed safety analysis (including TLAAs) and modifications, as well as after reassessment of safety related modifications for LTO. The period for making changes in the report and submitting it to the regulatory body depends on the national regulations, and this can range from immediately through annually.

The FSAR (including its supplements) and/or other licensing documents provide descriptions of activities in support of safe LTO to ensure that the operating organization maintains the necessary information to reflect the current status of the plant and addresses new issues as they arise.

The information to be included in the FSAR is described in Section 6.2 of Safety Reports Series No. 106, Ageing Management and Long Term Operation of Nuclear Power Plants: Data Management, Scope Setting, Plant Programmes and Documentation [16].

5.2. PLANT CONFIGURATION MANAGEMENT

Member States have general requirements for the licensees to keep updated records of plant configuration that are also appropriate for the LTO period. The basic requirements specified in SSR-2/2 (Rev. 1) [3] also apply to LTO, but the importance of plant configuration management is highlighted during the LTO process because of the modifications and refurbishments usually required as part of the process. Documentation prepared and submitted with the licence renewal application, with the LTO application or with the PSR is generally required to be consistent with the plant configuration and the design documentation. It is typically also required to keep track of modifications and other repairs, replacements and upgrades.

5.3. MANAGEMENT OF MODIFICATIONS

Typically, Member States have requirements for a programme for major modifications, reconstructions and replacements and for the management of modifications, either as a separate rule (e.g. in a guidance document) or as part of the overall regulations. Not all Member States regulate organizational and document modifications; some focus solely on physical modifications. Documentation modification management is usually required by the regulatory body as part of the plant's quality management system.

Review of all modifications of SSCs, changes of operational limits, conditions, instructions and procedures is usually part of a safety assessment or safety review and documented within these review processes. Many Member States have dedicated requirements established by the operating organization or by independent organizations (regulator, quality auditor, international peers) on configuration management and modification management. The effects of each safety important modification to the design basis of the plant and modified SSCs need to be adequately assessed. This ensures that the modifications are properly documented and retained in an auditable and retrievable form for LTO. The recommendations given in other IAEA documents (see Requirement 11 of SSR-2/2 (Rev. 1) [3] and para. 3.167 of IAEA Safety Standards Series No. SSG-61, Format and Content of the Safety Analysis Report for Nuclear Power Plants [17]) also apply.

There are opportunities for synergies between power uprate and licence extension. However, uprating will involve additional modifications, and these will also need to be assessed by the regulatory body. The impact of power uprate on ageing management for LTO is an important issue. Plant ageing issues may be aggravated by power uprate as a result of changing operating conditions. As with all changes to plant operating conditions, the effects of power uprate on the ageing of SSCs are assessed and enhancements to AMPs might be required. This may necessitate the installation of additional monitoring systems for certain critical components to ensure extended plant operational life [18]. Regarding power uprating, either the appropriate regulatory requirements are developed, or it is confirmed that the existing requirements provide sufficient guarantee for maintaining the safety level during the modifications, which could be competing with those for LTO in terms of decreasing conservativism in the safety analyses.

5.4. DESIGN BASIS DOCUMENTATION

The design requirements for the SSCs of nuclear power plants are established in SSR-2/1 (Rev. 1) [2]. During plant operation, compliance with the design basis is maintained according to the respective requirements. Compliance is typically comprehensively confirmed every 10 years during a PSR. When approaching LTO, it may be justified and requested by the regulatory body that the licensee completely review the design basis of the plant with the planned LTO period in view. Whatever method is used for approving LTO (e.g. a PSR, licence renewal, a separate licensing process or a combination of these), it is required that a full set of information about the design basis be available. The review of the plant's design basis can identify a key set of necessary documentation for the LTO programme (see para. 4.13 of SSG-48 [4] as an example regarding revalidation demand for TLAAs) and for ensuring safe LTO. The outcome of the review can show what data, documentation and analyses are required to be updated, supplemented or reconstituted to establish a sound basis for a safe LTO.

The experience of the Member States shows that, as an outcome of a PSR or a separate design basis review of an older nuclear power plant, discrepancies are revealed in many cases between the design documentation and the actual plant configuration. Some Member States reported that the documentation of the design basis of SSCs important to safety was incomplete, not sufficient or not obtained at all from the original vendors or designers. The outcome of the analysis may also reveal that the original design basis is less conservative or falls short of the safety level that is required by the regulatory body.

In cases where the design basis review reveals incompleteness, an effort needs to be made to retrieve or reconstitute the design basis. Reconstitution or re-establishment of the design basis is the responsibility of the operating organization for the preparation for a safe LTO (see para. 4.15 of SSG-48 [4]).

The reconstitution is to take into account the original design intent, the design philosophy and all the details of the implementation of the design, including any design modifications carried out during operation. In many

Member States, the design bases of nuclear power plants had to be extended over a certain set of accident conditions that had not been part of the original design. In some Member States, this was carried out as part of preparation for LTO.

Typically, the reconstituted design basis will be formulated based upon the outcome of a series of studies, engineering analyses, engineering judgements and R&D, all with a view of the planned duration of LTO. In this process, the regulatory body will typically require that all of the assessments and the prepared documentation be sufficiently detailed. The whole effort is a significant challenge for both the operating organization and the regulatory body.

The regulatory body and the operating organization need to agree on how to manage the reconstitution. This may include elaborating specific legal requirements and associated regulatory guides, specifying expectations for a separate methodology and/or the development of a dedicated review plan or inspection programme for keeping track of the reconstitution process. The implementation of a dedicated inspection programme can help the regulatory body receive up to date information and identify specific weak points of the process and take further steps (in due time), if necessary.

According to the experience of one Member State, during the collection of design basis information several difficulties were encountered. For example, basic data for equipment qualification were missing, as the vendor did not provide any information on the qualification of the concerned equipment. In this case, the qualified status was established with the help of research carried out by a contractor of the nuclear power plant. Another difficulty was the lack of fatigue calculations regarding safety class components, which the operating organization had to recalculate with the help of a contractor for safety class 1 and 2 components according to the expectations of the regulatory body. Because of the poor documentation of the design basis, further challenges were faced for specific TLAAs, since when the operating organization attempted to perform up to date strength calculations, they discovered that the calculations were mostly missing and only their final results were known. To provide the basic input data for these calculations, the operating organization performed a survey of the real dimensions of the safety class 1 and 2 mechanical components to validate the related drawings. A one-time inspection was conducted regarding the survey by the regulatory body, which, with a graded approach, focused on safety class 1 components.

In some Member States, the regulatory body uses INSAG-19 [19] as a reference document for defining requirements for design basis reconstitution for LTO as described in this section. According to Ref. [19], the regulatory body might require the establishment of a so called 'design authority', a formally designated entity within the operating organization, through formal arrangements, which may include the following approaches:

(a) Developing the complete design authority role within their own organizations and hence obtaining and maintaining all the information held by the original designers;
(b) Assigning the formal responsibility of 'responsible designer' to the original designers or their replacements;
(c) Defining a combination of these two approaches.

Irrespective of which formal arrangement is required by the regulatory body, the regulatory expectation for LTO is that the operating organization either obtains and maintains, in an auditable way, all the design information and documentation typically held by the designers, or has unlimited access to it. The outcomes of the design basis reconstitution are to be included and considered in the relevant plant programmes, and documented, including updating the FSAR. The regulatory body ensures that this process is carried out appropriately, and reviews and approves the modification of those documents and programmes that are subject to regulatory approval.

In those nuclear power plants for which the design basis information is well documented, and where information on all modification during its operation is incorporated into the design basis as part of the integrated management system of the operating organization, the review and provision of a sound design basis by the operating organization for a safe LTO involves a significantly reduced effort. Nevertheless, the regulatory body may request that the operating organization evaluate the cumulative effects of individual plant modifications during the original design life of the plant in view of the modifications proposed for LTO, in order to confirm that safety will not be adversely affected.

The final regulatory review process may take place during a PSR for LTO, licensing of LTO or in a separate regulatory process. The reconstituted design basis is used as the basis for any plant modification and other relevant safety assessments, including those after safety related events.

5.5. RECORD KEEPING AND DOCUMENTATION

Record keeping and documentation are general requirements (e.g. Requirement 35 of GSR Part 1 (Rev. 1) [1] for regulatory bodies and Requirement 15 in SSR-2/2 (Rev. 1) [3] for operating organizations). Some Member States regulate LTO by listing requirements for documentation to be updated and provided to the regulatory body for review and for its decision regarding licence renewal. Some Member States explicitly specify that documents and records be kept up to date, easily accessible (retrievable) and auditable specifically throughout the LTO period. The ability to review the design basis against updated requirements, documentation of the original design records, commissioning records and records of modifications is essential to validate the plant configuration for LTO. Thorough documentation of the LTO policy, programme and results, including the following, is typically required by the regulatory body in order to allow for an independent assessment of the condition of the nuclear power plant for safe LTO at any time:

(a)   The scope setting methodology, criteria and results;
(b)   Definition and establishment of commodity groups;
(c)   Ageing management and other plant programmes;
(d)   AMR methodology and results;
(e)   Identification and revalidation of TLAAs.

Further information on expectations for record keeping and documentation for the items listed above can be found in Section 2.2 of Ref. [16].

5.6.   PLANT PROGRAMMES

Maintenance, equipment qualification, ISI, surveillance and water chemistry are existing plant programmes that are essential for ageing management and evaluation for LTO [4]. Most Member States have established the requirements and guidance for these plant programmes.

Operating organizations are responsible for establishing these programmes and typically formal review and approval by the regulatory body will be required. However, in some Member States the operating organization will be required to establish these programmes or part of them according to the regulatory requirements, which define the format and contents for detailed review and approval before implementation.

These programmes receive special attention during the preparation for and implementation of LTO due to the importance of the management of ageing and degradation of in-scope SSCs. The effectiveness of these programmes in supporting LTO needs to be assured.

In the following subsections, the plant programmes are discussed following the sequence described in Section 4 of SSG-48 [4]. Repair and replacement are discussed in Section 5.6.1, and technological obsolescence is discussed as part of ageing management in Section 6.

Technical requirements are usually described in plant documents and industry standards. For LTO, the plant programmes are subjected to compliance reviews against the current national and/or international standards, codes, proven procedures and practices contained in SSR-2/2 (Rev. 1) [3] and Ref. [10]. Gaps are justified and agreed upon with the regulatory body. For example, the IGALL AMPs are structured according to the nine generic attributes of an effective AMP, in line with SSG-48 [4]. These attributes are considered essential for the plant programmes credited for LTO [8].

The plant programmes may also be subjected to ongoing compliance verification activities, including reviews, assessments, inspections and audits by the regulatory body, in addition to self-assessments by the plant and, in some cases, third party audits. All compliance activities are documented and record the objective evidence of compliance or conformance.

The plant programmes are interlinked as their outcomes are used for ageing management during LTO. This is done by adopting a systematic approach to taking corrective actions in an integrated manner.

5.6.1. Maintenance

Maintenance requirements are derived from design or reliability considerations. Regulatory documents outline the requirements for maintenance programmes for nuclear power plants, consistent with IAEA Safety Standards Series No. NS-G-2.6, Maintenance, Surveillance and In-service Inspection in Nuclear Power Plants [20], and operating experience feedback. The maintenance programme for a nuclear power plant usually needs to cover preventive and remedial measures, both administrative and technical. The measures are necessary to detect and mitigate the degradation of a functioning SSC or to restore the performance of the design functions of a failed SSC to an acceptable level during the original design life of the plant.

The requirements can specify that at the start of ageing management implementation, in-scope SSCs are included in the scope and can be categorized as active and passive components. The ageing management of active SSCs is partially covered by the plant maintenance programme. Regulatory oversight of the maintenance programme takes this into account.

The extent of the regulatory body's involvement in the review of maintenance activities during LTO depends on the practices in Member States. In general, the regulatory body verifies that such activities are properly conducted, particularly for SSCs important to safety.

The licensee's maintenance programme will typically be required to specify the link(s) with the ageing management programme, including the frequency of maintenance activities and specific information to be exchanged between different plant programmes. Specific tagging of ageing management activities in the maintenance management systems is also needed. The consistency of the maintenance programme activities for ageing management are to be verified against the generic attributes of an effective AMP.

In some Member States, the maintenance rule and ageing management will require that the performance of certain SSCs be monitored, to provide assurance that they are capable of performing their intended functions. Generally, this requirement for monitoring will continue in the LTO period. Further, all SSCs classified as not important to safety, but whose failure could prevent satisfactory accomplishment of any of the safety functions, are also expected to be reviewed as part of these programmes.

The maintenance activities for and during LTO may include review of maintenance programme activities for monitoring, surveillance, inspection, testing, assessment, calibration, work management, preventive and corrective maintenance, health monitoring of systems and components, service, overhaul, repair and replacement of parts.

The regulatory body will generally require the licensee to have an efficient maintenance management process in place to keep the backlog for corrective maintenance tasks reasonably low throughout the intended period of operation. The results of all maintenance activities may be fed back through an optimization process that enables continuous improvement of the programme. The regulatory body may use the safety performance indicators reported by the licensee to trend safety performance and effectiveness of maintenance during LTO, in line with IAEA Safety Standards Series No. SSG-50, Operating Experience Feedback for Nuclear Installations [21].

Compliance verification inspections can be carried out periodically to confirm maintenance programme effectiveness and compliance with regulatory requirements.

An example of a requirement for maintenance effectiveness monitoring can be found in Ref. [22].

Repair and replacement activities are carried out as part of the maintenance programme during the original design life and LTO. It may be necessary to repair or replace some SSCs as a result of condition assessments for LTO. The implementation of corrective actions as a result of repair and replacement activities may be performed on a routine basis in accordance with applicable nuclear and other industrial codes and standards (e.g. ASME, IEC, IEEE, ISO) and the respective acceptance criteria. Repair and replacements for SSCs important to safety are performed after safety review and analysis and approval by the regulatory body, as applicable. The repair and replacement activities have links with AMPs, and the specific information to be exchanged between these programmes is determined.

5.6.2. Equipment qualification

In most Member States, regulatory requirements and guidance exist for equipment qualification that cover environmental, seismic and electromagnetic and radio frequency interference aspects relevant for the given SSCs. Accordingly, the regulatory body might require the operating organization to establish, implement and maintain equipment qualification programmes, such as environmental qualification, seismic qualification and electro-magnetic qualification. Equipment qualification demonstrates that, throughout its qualified life, the equipment will still be capable of performing its intended function(s) under the full range of specified service conditions, for the mission time, including the most severe environmental conditions during design basis accidents. Environmental monitoring includes measurement of stressors, such as temperature, radiation and relative humidity. Further, other service parameters related to qualification can be monitored, such as operational cycles during normal operating conditions, and results of periodic testing. The operating organization develops and maintains the master list of the SSCs requiring equipment qualification, which can be reviewed by the regulatory body during the authorization for LTO. Any changes to equipment qualification that impact the licensing basis for LTO are subject to prior approval of the regulatory body through a formal modification process. In some Member States, SSC qualification and environmental conditions are assessed annually, while for LTO they are assessed through a PSR or through a separate review.

The regulatory body will typically require the operating organization to reassess or confirm the qualification of equipment for LTO [4], by either test or analysis or a combination of both, in line with Requirement 13 of SSR-2/2 (Rev. 1) [3]. The operating organization needs to verify the consistency of the programme with the generic attributes of effective AMP (see para. 4.17 of SSG-48 [4]).

The regulatory body stipulates that the effectiveness of the equipment qualification programme is reviewed periodically by the licensee and that the scope and details of the equipment qualification process are documented and submitted for regulatory review and/or approval.

5.6.3. In-service inspection

The SSCs are examined for possible defect indications to determine whether they are acceptable for continued safe operation or whether remedial measures are to be taken. Emphasis is placed on examination of the pressure boundaries of the primary and secondary coolant systems and the containment because of their importance to safety and the potentially severe consequences of their failure. The regulatory body typically requires the operating organization to establish and implement an ISI programme, consistent with industrial codes and standards, for plant operation during the original design life of the plant. This programme is assessed and enhanced, as necessary, to assure adequate ageing management and evaluations for LTO of applicable in-scope SSCs, including consideration of baseline data, inspection results, AMR and maintenance activities. The regulatory body might require the operating organization to extend the ISI programme with subsequent inspections for LTO and submit it for regulatory review and approval.

The regulatory body may specify that a database for activities be developed and maintained to verify the adequacy of in-service examination in detecting, characterizing and monitoring the degradation of structures or components and to support the findings and the conclusions necessary for ageing management decisions [4]. The regulatory body typically requires the operating organization to verify and check the consistency of the ISI programme with the generic attributes of an effective AMP. During the collection of feedback from ISI for AMR, the regulatory body reviews the results and degradation mechanisms revealed during ISI. The regulatory body ensures that these results are taken into consideration for and during LTO.

5.6.4. Surveillance

The regulatory body makes sure that the operating organization has established and implemented adequate surveillance programmes, including functional tests, as a basis for ageing management and evaluations of applicable in-scope SSCs for the LTO. The surveillance programmes address the requirement for the safety margins for LTO to be adequate and provide a high tolerance for anticipated operational occurrences, errors and malfunctions.

Performance characteristics that are inconsistent with those assumed in the safety analysis may be identified as a result of surveillance activities and be both corrected and considered in the regulatory processes. Test requirements can be re-evaluated after modification(s) or based on relevant operating experience. Regulatory requirements may be confirmed through the following activities:

(a) Governing documents are reviewed to assess compliance with the applicable standards when such documents are revised to reflect updates to the standards.
(b) Documents from manufactures and suppliers are reviewed to verify component lifetimes.
(c) A chronological record of the SSCs' operation is reviewed to confirm whether a component has been operated within operating and/or design limits.

(d) Outage inspection reports are reviewed to assess compliance with the processes described in the programmes.

(e) Compliance verification inspections are carried out periodically to confirm compliance with requirements that are difficult to verify through document reviews.

Material surveillance programmes, such as the reactor vessel material surveillance programme, are reviewed and can be extended or supplemented for ageing monitoring within the period of LTO, if necessary.

The surveillance programmes of nuclear power plants need to be checked for consistency with the nine generic attributes of an effective AMP (see SSG-48, table 2). The regulatory body might require an operating organization to extend the surveillance programmes for LTO and submit them for review and approval.

5.6.5. Water chemistry

Through appropriate requirements, the regulatory body ensures that the operating organization has established a water chemistry programme with information about the chemical and radiochemical environment for the integrity of structures or components within the scope of ageing management and evaluations for LTO. The water chemistry programme specifies processes, specifications, overall requirements, parameter monitoring, data trending and evaluation to ensure effective control of plant chemistry during operational and lay-up conditions. Long term effects arising from operational and environmental conditions during operation can be evaluated, assessed and monitored as part of the chemistry programme for LTO.

In some Member States, the regulatory body will require the operating organization to establish a water chemistry programme and chemistry performance index to make comparisons between the concentration of selected impurities and corrosion products and corresponding limiting values established during the operation of the plant, to ensure safety for regulatory review and approval during the LTO. The water chemistry programme specifies its link with other AMPs, and the specific information to be exchanged between these programmes and documented for verification. Compliance verification inspections can be carried out periodically to confirm chemistry programme effectiveness and compliance with requirements. The water chemistry programme needs to be checked for consistency with the nine generic attributes of an effective AMP.

5.6.6. Corrective action programme

The regulatory body typically requires that the corrective action programme ensures the necessary enhancement of the AMPs and other plant programmes to manage the ageing effects and improve the overall safety of the nuclear power plant for LTO. In preparation for LTO, corrective action tasks are typically completed and plant programmes updated where necessary. The regulatory body might require new corrective actions to be implemented within a specified time frame in preparation for and during LTO.

5.7. ADDITIONAL PLANT PROGRAMMES

In some Member States, programmes in addition to those mentioned in Section 5.6 are implemented and might be required by the regulatory body to be updated and submitted for regulatory review and approval for LTO. These may include programmes for radiation protection, radioactive waste management, emergency preparedness and response, environmental monitoring and quality assurance, and programmes covered under technical specifications. The objective is to identify changes needed and possible improvements to these programmes to ensure and enhance the plant's ability to meet the design and safety criteria and licensing basis for the LTO period.

6. NATIONAL REQUIREMENTS AND GUIDES FOR AGEING MANAGEMENT FOR LONG TERM OPERATION

This section describes the regulatory requirements and guidance to be adopted in the Member States for the regulation of ageing management.

6.1. ORGANIZATIONAL ARRANGEMENTS

In some Member States, the regulatory body might require the review and/or set up of an organizational arrangement in the plants for LTO preparation and implementation, and/or an authorized organizational entity (assigned by the operating organization) with responsibilities for ageing management and/or the LTO programme (see paras 7.3 and 7.4 of SSG-48 [4]).

The regulatory body may request that the plant create interdisciplinary ageing management teams consisting of members of different units of the plant and external experts, on a permanent or ad hoc basis, to ensure a comprehensive approach to AMP. Regulatory provisions may also address the need for external organizations to provide expert support services on specific topics, such as condition assessments and R&D.

The requirements for organizational arrangements also cover the provision of appropriate resources, including human resources, for the preparation and duration of LTO. The regulatory body assesses, inspects and approves the safety related modifications of organizations, in accordance with national regulatory requirements.

The role of plant personnel involved in establishing, implementing and assessing plant programmes is important. Training on the effect of ageing on SSCs for personnel involved in operation, maintenance and engineering to enable them to make an informed and effective contribution to ageing management may be emphasized. The plant's management may also consider a system to exchange information, inspection results and data within relevant organizations and to identify and resolve common occurrences in ageing management.

6.2. SCOPE SETTING FOR LONG TERM OPERATION

For safe LTO preparation, requirements for a well defined scope setting process are essential, in addition to appropriate guidance on the methodology for determining the scope. The IAEA's scope setting recommendations for ageing management for LTO can be found in Section 7 of SSG-48 [4]. Further information on scope setting is provided in Section 3 of Ref. [16]. Scope setting is a systematic process that covers all SSCs (see paras 5.14–5.21 of SSG-48 [4]):

The initial or master list of SSCs contains all items in the plant, from which SSCs in the scope of ageing management and LTO will generally be required to be selected according to the above considerations. In some Member States, the national regulations and guidance contain specific provisions on failures of SSCs that are not classified as important to safety but may prevent SSCs from fulfilling their safety functions; for example, fire protection piping that leads to the electric failure of an electrical panel that controls the current to a motor operated valve performing a safety engineered function. SSCs credited to function in design extension conditions are also considered in the scope setting process, as specified in national regulatory requirements. National requirements and guidance can take into account the detailed considerations for scope setting in Ref. [16].

The ageing of SSCs may increase the probability of common cause failures that could result in the impairment of one or more levels of the protection provided by applying the defence in depth concept. Therefore, in the requirements for setting the scope of ageing management for SSCs no credit is taken for redundancy among them.

In some Member States, the exclusion of SSCs from the scope is allowed if periodic replacement or scheduled refurbishment is planned based on predefined rules, taking into account designer and manufacturer recommendations or other bases, and not based on an assessment of the condition of the SSCs. This is in accordance with para. 5.17(a) of SSG-48 [4]. In some Member States, the regulatory bodies have specific provisions to extend the scope with additional SSCs, considering not only nuclear safety, but other safety related aspects, for example labour safety, industrial safety, conventional environmental protection or fire protection of SSCs not in scope.

The CLB typically includes an analysis of the safety relevance of all SSCs. The scope setting process for ageing management and LTO is based on, but not limited to, the existing safety classification of SSCs.

Documentation of the scope setting process varies between Member States [16]. In most Member States, system and component specific analyses will be required for a PSR and for LTO preparation.

Documentation of the scope setting process, specifically an assessment for all SSCs with a justification for including and excluding them, facilitates the review at the time of authorization and for subsequent verification. The requirement for the presentation of the scope may include, for example, schematic drawings, tables and databases. Determination of system and component boundaries is an important part of the regulatory requirements. Clarification of boundaries between the technical areas and boundaries of in-scope and out of scope SSCs is generally required by the regulatory body. Use of visualization tools for presenting the scope and boundaries may also be part of the requirements. An important aspect of the scope setting process is to provide a suitable procedure to demonstrate the completeness of the scope. In Member States, dedicated plant walkdowns are required as part of developing, verifying and achieving completeness; in particular, the identification of SSCs whose failure may jeopardize a safety function needs careful attention during these walkdowns. In addition, walkdowns might be required to confirm the correctness of the plant documentation, or as a complementary tool if there is no assurance that plant documentation is sufficiently accurate.

After scope setting, most Member States allow the grouping of SSCs into commodity groups according to similarities in properties, functions, materials and operating environment. Detailed guidance on the methodology of the grouping process and criteria may be provided by the regulatory body in such cases. There are two examples among Member States for such grouping:

(1) Commodity grouping is understood as the joint consideration of elements with similar characteristics (e.g. material of construction, operating environment, applicable ageing effects) or properties that justify their consideration as a single group in future ageing phenomena analysis;
(2) Grouping by systems that follows the natural organization of the components considered at the nuclear power plant.

Further information on commodity groups can be found in Ref. [16].

6.3. AGEING MANAGEMENT PROGRAMMES

The regulatory bodies verify that ageing management consists of design, operations and maintenance actions to prevent or control, within acceptable limits, the ageing of SSCs. Ageing management is an interdisciplinary activity that involves engineering, maintenance, surveillance, equipment qualification, ISI and safety analyses. Regulatory requirements typically specify that AMPs are developed using a structured methodology to verify that a consistent approach is adopted in implementing ageing management (see SSG-48 [4]) during the design life of plant operation. Long term effects arising from operational and environmental conditions (e.g. temperature conditions, radiation conditions, corrosion effects or other degradation mechanisms in the plant that may affect the long term reliability of plant SSCs) may be evaluated, assessed and monitored as part of ageing management (Requirement 14 of SSR-2/1 (Rev. 1) [2]) for LTO.

AMPs are developed according to the regulatory requirements and, in some Member States, submitted for formal regulatory review and approval before implementation. The AMP framework will provide for a comprehensive, overarching programme, or, alternatively, a 'road map' document that demonstrates existing processes and programmes for effective ageing management. A typical set of AMPs for the most common technologies can be found in Ref. [8].

AMPs typically follow the recommendations provided in Section 5 of SSG-48 [4], especially to be consistent with the nine generic attributes of an effective AMP (see table 2 [4]).

AMPs include the ageing management of mechanical, electrical, instrumentation and control (I&C), and civil structures and components. In some Member States, regulatory review and approval for ageing management are done through a PSR to support the decision making process for LTO, which is usually performed at 10 year intervals. The regulatory body carries out reviews and inspections of AMPs.

6.4. AGEING MANAGEMENT REVIEW

In most Member States, ageing management for SSCs is included in operation, inspection and maintenance activities, and its review is part of a PSR or similar assessment. As a minimum, ageing management needs to be subject to periodic review during each PSR, but typically it is reviewed more frequently.

Many SSCs are subject to some form of physical change caused by ageing, which could eventually impair their intended functions and reduce their service life. In order for the licensee to be able to demonstrate that ageing of the in-scope SSCs is being managed effectively, an AMR will typically be required to be conducted on all in-scope SSCs (see paras 5.22–5.36 of SSG-48 [4]). Scope setting is an important part of the review, depending on the approach of the operating organization, and as a consequence the scope can be modified significantly from the start of the operation until the start of LTO. The AMR may also be performed via commodity groups of structures or components (see para. 5.20 of SSG-48 [4]). The AMR ensures that all ageing effects requiring management are identified for each SSC or commodity group of structures or components. The AMR is not to be confused with the review of the overall or plant level AMP.

AMR is a systematic identification and assessment of degradation mechanisms and related ageing effects, and confirmation that all of them are managed by effective plant programmes or TLAAs. The AMR is performed by the licensee for all in-scope SSCs to ensure that each one fulfils its intended functions considering its current and forecasted condition. In addition to analysis of plant documentation, plant walkdowns are suggested to be performed to verify the condition of the SSCs.

As part of preparation for LTO, the AMR is expected to be comprehensive and take into account relevant information arising from design, fabrication and operation and their combined effects for the LTO period of the plant.

According to paras 5.25 and 5.26 of SSG-48 [4], a process is needed to identify relevant ageing effects, degradation mechanisms, environment and stressors for each structure or component, and to ensure that programmes to manage the identified ageing effects and degradation mechanisms are in place. This process consists of the following steps (see para. 5.25 of SSG-48 [4]):

(a) Review of the list of SSCs within the scope of LTO;
(b) Assessment of the current condition of the structure and component;

(c)   Identification if a TLAA is applicable (e.g. does a TLAA exist?);

(d)   Identification of relevant ageing effects and degradation mechanisms;

(e)   Verification that existing AMPs or other plant programmes that manage the ageing of structures or components are consistent with the nine generic attributes of an effective AMP;

(f)   Review of the effectiveness of the AMPs;

(g)   Development of new programmes or modification of existing programmes;

(h)   Verification of the qualified lifetime of equipment important to safety and necessary corrective actions in accordance with the equipment qualification programme.

The expected result is an AMR report demonstrating the effective management of ageing effects and degradation mechanisms for each in-scope SSC.

Following discussions with the operating organization, the regulatory body may issue step by step guidance on the process of AMR to facilitate licensees conducting a successful review starting with scope setting (and screening, where appropriate, taking into account para. 5.17 of SSG-48 [4]) until its results can be documented.

Relevant data from design, manufacturing, commissioning, operation and maintenance history, tests, inspections and monitoring are collected for each SSC or commodity group. All information and conclusions with regard to the scope of AMR are then expected to be documented according to para. 5.34 of SSG-48 [4]. The results of AMR may include but are not limited to the following items:

(a)   A list containing information on the applicable degradation mechanisms of the in-scope SSCs and the appropriate programmes and further actions to be implemented to preserve their intended functions;

(b)   The current performance and condition of SSCs, including assessment of any indication of significant ageing effects;

(c)   A proper justification from the operating organization for those SSCs and environments where ageing mechanisms are not considered to be applicable; the regulatory body expects a proper justification from the operating organization.

Applicable combinations of material, environment and ageing effects are expected to be managed by an AMP, which may involve modification to assure effectiveness, or other plant programmes, including the development of a new AMP, if necessary.

The IGALL master table [8] can be used to confirm that the in-scope SSCs are subject to adequate ageing management. It can further be used to confirm consistency between the operating organization's approach to ageing management and IGALL.

6.5. REVALIDATION OF TIME LIMITED AGEING ANALYSES

The programme for LTO typically includes the revalidation of TLAAs, if necessary; hence the intended function(s) of the SSCs in scope for TLAAs will be maintained throughout the planned period of LTO consistent with the licensing basis, in accordance with SSG-48 [4].

The regulatory body will typically require revalidation of the TLAAs by the operating organization before LTO, according to para. 7.28 of SSG-48 [4]. The regulatory body may require a proactive revalidation process for TLAAs in advance of considering the condition of the SSCs at the end of the planned period of operation. The rationale for such projection is a key element that needs to be proposed by the plant and accepted by the regulatory body, prior to the beginning of the revalidation work. This needs to be done well in advance, otherwise an unsatisfactory outcome of the TLAA revalidation work done by the operating organization may cause delay in the licensing activity.

In some Member States, an additional safety margin is required for revalidation of TLAAs (and performance of AMR). For example, in one Member State a margin of 10 years was required, that is, the demonstration was to be performed for an additional 10 year period compared to the intended duration of LTO.

Within the framework of LTO, the complete identification and revalidation of all relevant TLAAs for in-scope SSCs is critical, since these analyses are not only used to demonstrate safe plant operation, but may also in many cases have an impact on other activities, such as the scope or type of ISI. As an example, fatigue TLAAs can be used to reduce or extend the scope of needed ISIs.

Based on experience feedback, some Member States have faced difficulties during the assessment of the TLAAs due, for example, to an incomplete list of applicable TLAAs, unclear scope of TLAAs, and non-uniform methodology and/or structure for the analysis. This has caused the review to be more resource intensive than was anticipated. To avoid these difficulties, the regulatory body can issue guidance containing a list of potential TLAAs and step by step instructions for TLAA identification and the TLAA revalidation methodology. In order to develop this guidance, the regulatory body may need to perform its own research, possibly involving TSO(s). The list of TLAAs from the IGALL programme [8] can serve as a benchmark that may be taken into account in national regulations and in regulatory guides. If the regulatory body issues guidance on the conduct of the review and/or revalidation of TLAAs, it is possible to compare this list with the TLAAs identified by the operating organization or by the original vendor. Depending upon the plant technology, there may be a need to identify additional plant specific TLAAs for specific systems and components. Review of the FSAR and its background documentation by the operating organization can be used to identify any time limited assumptions made by designers. Another possibility is the implementation of international benchmarking, which was reported by some Member States. During this benchmarking, international experts reviewed the revalidation and/or development work done by the operating organization and its TSO(s) and provided their experience and suggestions. If some potential TLAAs are not applicable as identified by the plant, the rationale for such a conclusion needs to be documented and submitted to the regulatory body for further discussion.

If the outcomes of TLAAs do not fulfil the acceptance criteria established in the design basis or safety analysis of the plant and cannot cover the whole LTO period for some components, appropriate corrective actions need to be taken by the operating organization, as described in para. 5.68 of SSG-48 [4], where three acceptable corrective action types are listed. The first possibility for corrective actions can be modification, repair, or replacement of the concerned equipment, based on the decision of the operating organization. This decision needs to be established and communicated effectively to the regulatory body as per the framework of the licensing activity. Another possibility is the refinement of analysis to decrease existing conservatism, which needs to be communicated to the regulatory body, whose approval is required. It is also possible to implement further actions in operation, or manage the problem that has arisen with maintenance, ageing management or other plant programmes. These corrective actions are inputs for development of the LTO programme. Some Member States review the proposed actions in conjunction with the PSR results and following the global assessment methodology, as described in SSG-25 [7].

As a prerequisite of LTO, the regulatory body typically requires that TLAAs be kept valid until the end of the planned LTO period so that the operating organization will have sufficient provisions to ensure that any further modifications in plant configuration will not jeopardize the licensing basis.

6.6. MANAGEMENT OF TECHNOLOGICAL OBSOLESCENCE

In some Member States, the regulatory body requires the licensee to establish and implement a technological obsolescence programme (TOP) to address all SSCs important to safety and manage the spare parts required to maintain those SSCs during the design life and for the LTO of the plant. Such a TOP may involve the participation of the engineering, maintenance, operations and work planning units, senior plant management and supply chain organizations. The TOP is submitted to the regulatory body for review and assessment, as applicable.

7. PERIODIC SAFETY REVIEW WITH RESPECT TO LONG TERM OPERATION

This section summarizes the regulatory approach of the Member States using the PSR methodology for LTO.

According to Requirement 12 of SSR-2/2 (Rev. 1) [3], PSR is a systematic reassessment of the safety of an operating nuclear power plant carried out at regular intervals to assess the cumulative effects of ageing, modifications, equipment requalification, operating experience, technical developments and siting aspects, national and international safety standard updates, and organizational and management issues. PSR aims to ensure safety throughout the service life of the facility. Service life is the period from initial operation to final withdrawal from service and includes LTO. Recommendations on how to conduct a PSR are provided in SSG-25 [7].

Most of the Member States have requirements and guidance for conducting PSRs in accordance with Requirement 32 of GSR Part 1 (Rev. 1) [1], and some use the PSR as part of the LTO authorization process. Member States that are also members of the European Union are required to follow Issue P in Refs [23] and [24] for conducting a PSR at least every 10 years.

Some Member States consider a PSR to be a prerequisite for LTO, as it provides an effective way to obtain an overall view of actual plant condition with respect to safety, allows the regulatory bodies to detect issues of concern to be addressed for supporting safe operation during the period of LTO, and also identifies areas for safety improvements. Where applicable, the outcome of a PSR is an important input for the decision making process by the regulatory body to approve LTO.

In addition to routine continuous safety improvements, some regulatory bodies, within their regulatory framework, use the opportunity of LTO to increase the plant's safety margins beyond their current level, and to apply improvements in technology or plant safety level identified from the requirements of modern standards and internationally recognized good practices. As an example, in some Member States safety improvements are required to mitigate design extension conditions within the scope of LTO.

The latest PSR results prior to entering LTO (used to justify LTO; hereafter PSR for LTO) can be more extensive in scope than the previous PSRs.

The basis document for the PSR for LTO is the starting document to address all the requirements for LTO in accordance with the national regulatory framework and to manage the scope of each of the safety factors. Therefore, defining the requirements for the format and contents of the basis document is of key importance. For example, assessment of safety factors 1 (plant design),

2 (actual condition of SSCs important to safety), 3 (equipment qualification) and 4 (ageing) will form the core part of the plant's licensing basis for the extended period of operation; hence, these areas are covered in more detail in the requirements. The requirements for these safety factors include consideration of the feedback coming from the assessment of safety factors 8 (safety performance) and 9 (use of experience from other plants and research findings), as they provide valuable information for safety improvements.

The intended period of LTO needs to be defined clearly within the basis document, and sound scope setting is to be performed and documented in accordance with SSG-48 [4]. In addition to the licensing basis, there is a need for the regulatory body to define cut-off dates to be used in plant safety factor review, as these may influence not only the editions of standards and codes to be considered, but also the time frame for operating experience feedback considered in the PSR.

The process for the development of the basis document depends on the different Member States' legislative and regulatory frameworks (see Section 2 for more details). In some Member States, the current legislation covers LTO expectations and rules and the basis document is developed by the operating organization and submitted to the regulatory body for review and approval. In others, the regulatory body is involved in the development process for the basis document at an early stage, with the purpose of making the expectations available to the plant and arriving at a common agreement.

In the case of safety factor 1 (plant design), the regulatory body may need to supplement the PSR requirements based on confirmation of the cumulative effect of all of the modifications made to the original plant design for any new proposals arising from different sources. Such sources are typically the comparison of the current design against modern standards, new information from operating experience, and backfitting after new regulatory requirements are promulgated.

In addition, the regulatory body needs to set out clear requirements for safety factor 2 (actual condition of SSCs important to safety) to make sure that degradation mechanisms are comprehensively identified, and ageing effects will continue to be identified and managed for each structure or component in the scope of LTO for the planned period of operation.

Regarding safety factors 3 (equipment qualification) and 4 (ageing), as with other plant programmes, the key requirement to be established by the regulatory body is related to the effectiveness of the AMPs through the fulfilment of the nine generic attributes as defined in SSG-48 [4].

All of the above mentioned considerations and PSR safety factors are a necessary but not sufficient input for a comprehensive and sound LTO programme as specified in Requirement 16 of SSR-2/2 (Rev. 1) [3]. Consequently, the supplements are provided in the requirements for PSR for LTO compared to a regular PSR with regard to LTO preconditions, degradation mechanisms and ageing effects, and revalidation of TLAAs and AMR, taking into account all major aspects according to SSG-48 [4]. As an example, the Belgian regulatory approach mentioned that revalidation of TLAAs and AMR are specifically within the scope of a PSR.

Based on the findings of the PSR and considering the global assessment (see Section 6 of SSG-25 [7]), the regulatory body expects an integrated implementation plan for proposed safety improvements (including their safety significance and prioritization) from the operating organization, in line with para. 8.19 of SSG-25 [7]. In some Member States, the operating organization submits the methodology used for assessing, categorizing and prioritizing safety improvements to address findings for further regulatory review and approval. According to SSG-25 [7], the approach adopted in the global assessment could be based on deterministic analysis, probabilistic safety analysis, engineering judgement, cost–benefit analysis and/or risk analysis, or a combination thereof. Some regulatory bodies specify the appropriate requirements for the global assessment accordingly.

Considering that the implementation of safety improvements and plant modifications indicated from the PSR for LTO might affect the original design intent (it might have negative effects if the interaction of the safety improvement with the SSCs, as well as the cumulative effects of all plant modifications, are not carefully assessed), it is a justified practice for the operating organization to consider assessing the adequacy of the plant's defence in depth concept for the LTO, as derived from the proposed integrated implementation plan.

Some Member States assess the defence in depth capabilities, including both the design features and the operational measures taken to ensure safety, as well as the provisions implemented during any stage of the lifetime of the plant, based on the methodology described in Safety Reports Series No. 46, Assessment of Defence in Depth for Nuclear Power Plants [25].

The requirements for PSR for LTO cover the approach to address corrective actions and any improvement in plant safety level for LTO and to demonstrate that the remaining risk will be kept as low as reasonably practicable.

8. PREPARATION OF THE REGULATORY BODY FOR LONG TERM OPERATION PROGRAMME REVIEW

This section describes the practices of the regulatory bodies in Member States in preparing for the oversight of the licensees' LTO programme.

The regulatory body's review and assessment represent one of the core regulatory functions [1]. This interacts with other core regulatory functions and processes, particularly the development and/or provision of regulations and guides, authorization, regulatory inspections and enforcement. The review and assessment of relevant information by the regulatory body are general requirements (Requirement 25 of GSR Part 1 (Rev. 1) [1]).

The review and assessment of information for the LTO programme are carried out as part of the authorization process. The objective is to ensure that all safety requirements are addressed, and the prescribed level of safety is assured. The review and assessment process is sufficiently flexible to allow its modification for checking compliance with regulatory requirements and to facilitate the use of findings of relevant studies and routine or special safety reviews and inspections. The review and assessment process is conducted differently in accordance with national practices. The process and its results need to be well documented.

The regulatory review and assessment of the LTO programme can be divided into several major steps:

(a)   Preparation for the review and assessment;
(b)   Conduct of the review and assessment;
(c)   Analysis and resolution of the findings, determination of safety improvements and decision on the acceptability (approval) of the operating organization's safety argument for LTO;
(d)   Reporting and documentation and preparation of an evaluation report by the regulatory body;
(e)   Review of the implementation of the LTO programme.

The main regulatory activities identified in preparation for the LTO programme review and assessment are summarized in this section, including the regulatory body responsibilities and tasks. This section consists of five subsections dedicated to the specific activities covering general considerations: the development and provision of regulations and guides; the preparation of staff (training); the development of internal regulatory processes; preparation for LTO programme oversight, including review and assessment and licensing; and preparation for the use of results from PSR (or similar systematic assessments) for LTO.

The activities of the regulatory body for evaluation of the plant documentation and programmes relevant to LTO, the assessment, approval and implementation of the LTO programme, and reporting and documentation are described in Sections 9 and 10.

8.1. GENERAL CONSIDERATIONS

The establishment and implementation of a management system for the regulatory body is a general requirement (Requirement 19 of GSR Part 1 (Rev. 1) [1]). The management system is designed to effectively discharge the tasks and responsibilities related to the regulation, licensing, review and assessment, inspection and enforcement and all other activities for effective control of nuclear safety of facilities and activities by the regulatory body. The regulatory system incorporates the national legislation and obligations, recommendations and expectations of international agreements, standards and directives applicable to the regulatory activities in the area of nuclear safety. The regulatory body needs to confirm that existing management systems are adequate for the regulatory review of applications for LTO and oversight during LTO.

Usually, the regulatory body starts preparing for the LTO authorization process after announcement of the licensee's interest in the continuation of the plant operation beyond the originally considered operational lifetime and submission of the feasibility study for LTO to the regulatory body (e.g. Armenia, the Czech Republic, Slovakia). The feasibility study typically describes the benefits of LTO for the operating organization and is available several years before entering into LTO. The study can be prepared for LTO related activities by the operating organization. Such a study is a legal requirement appearing in the legislation of some Member States (e.g. Armenia). This arrangement ensures an opportunity for timely preparation of the regulatory body.

In the preparation phase the necessary legal basis may be created or supplemented to provide compatibility for LTO authorization (see Section 8.2).

The operating organization may also submit an LTO programme to the regulatory body according to Section 4.2. A typical time frame for the LTO programme to be received by the regulatory body is three to seven years before the LTO period commences.

The review and assessment process is subject to a graded approach (Requirement 26 of GSR Part 1 (Rev. 1) [1], para. 2.5(c) of GSG-13 [5]), so that the degree of review and assessment for the LTO programme(s) is commensurate with the magnitude of the possible radiation risks arising from the facility and activity. Factors that need to be considered in the preparation of regulatory activities include, for example, the history of the plant with respect to safety (major and less significant events, safety culture evolution, etc.), the maturity and complexity of the plant activity related to LTO, and the knowledge and expertise of the operating organization.

The regulatory body develops management projects, review plans or any other forms of regulatory documents within its management system to manage the LTO programme review and assessment and inspection. The regulatory body's internal documents may include:

(a) Definition of the scope of review and assessment and inspection process;
(b) Specification of the purpose of the review and assessment and inspections;
(c) Specification of the technical basis and identification of the methodology and criteria for the review and assessment and for the inspection programmes;
(d) Specification of the time frame and schedule for the review and assessment and inspections;
(e) Allocation of resources and specification and assignment of the roles and responsibilities of the review and inspection teams, and appointment of the review manager;
(f) Specification of the form of outputs (e.g. evaluation reports, reporting of findings, inspection reports).

When specifying the safety objectives and requirements to be used for the review and assessment and inspections, the regulatory body considers national laws and regulations, guidance documents and standards issued by national and international organizations, advice obtained from external experts and the advisory bodies of the regulatory body, applicable operating experience feedback and the results of R&D, and expertise used by others involved in reviewing and assessing similar facilities and activities with respect to safety.

The regulatory body plans the resources (both human and financial) for the review and assessment and inspection. Determination of allocated resources is dependent on the scope, organizational aspects and planned schedule for the review, and may consider a need to employ external organization(s), and training for regulatory body staff involved in the LTO programme review and assessment and inspection.

The planning ensures that qualified and competent staff of the regulatory body undertake the review and assessment or evaluate the assessment, if any, conducted by external organization(s) (Requirements 18 and 20 of GSR Part 1 (Rev. 1) [1]). The regulatory body does not rely solely on the safety assessments conducted by the operating organization. The regulatory body needs to be able to independently review the information submitted by the operating organization or information that comes from inspections and feedback on operating experience (Requirement 25 of GSR Part 1 (Rev. 1) [1]). In particular, with regard to LTO, the regulatory body reviews and assesses the LTO programme, the scope setting methodology and the scope of LTO, plant programmes relevant to LTO, AMRs, AMPs, TLAAs and other aspects of LTO specified in the legal and regulatory requirements. In addition to review and assessment, some regulatory bodies or their TSOs have performed verification calculations for TLAAs and other analyses for LTO decision making. This type of verification usually uses a graded approach to confirm the correctness of the most important methodologies, analyses and results.

Member States that have limited human resources may employ an internal review scheme that uses external support and international cooperation with other Member States' regulatory bodies or the review and support services of the IAEA, for example the peer review of safety aspects of long term operation (SALTO). In some Member States, external support is used, especially for reviewing TLAA submissions, to provide a review depth and expert assessment commensurate with the complexity of the analyses. GSG-12 [14] contains special considerations for avoiding conflicts of interest when a regulatory body uses external experts. Some regulatory bodies stipulate that the operating organization invite a SALTO mission to review preparedness for LTO and use the results in the regulatory decision making process.

Dedicated communication arrangements between the regulatory body and the operating organization are typically used for the LTO programme review and assessment (e.g. the appointment of contact persons, agreement on modes of communication and regular information exchange, meetings). This communication helps the regulatory body keep track of progress in the preparation and implementation of the LTO programme by the operating organization and with clarification of the regulatory requirements, when needed (Requirement 21 of GSR Part 1 (Rev. 1) [1]).

Interactions between the regulatory body and the operating organization, and the regulatory body and interested parties (e.g. public), can take place at different stages of the LTO review, depending on the practices of Member States. Communication and consultation with interested parties (Requirement 36 of GSR Part 1 (Rev. 1) [1]) may be important for the regulatory body to both inform and to recognize the view of interested parties regarding the operating organization's LTO.

Depending on national practice, the regulatory body may have responsibility for the following:

(a) Communicating with the operating organization regarding a feasibility study for an LTO programme (if prepared and submitted to the regulatory body and if it is part of the regulatory body's involvement in the LTO);
(b) Developing, initiating, specifying or approving the requirements for LTO;
(c) Developing and publishing regulatory guidelines for the LTO programme and its implementation;
(d) Informing the operating organization regarding the time schedule for conducting LTO oversight and the expected forms of the regulatory body outputs from the oversight;

(e) Approving the general scope, content and form of the documentation to be provided by the operating organization prior to submission of the LTO programme to the regulatory body or information accessible for LTO programme oversight and expected outcomes;

(f) Approving the concept, methodology and approach used for the LTO programme and the operating organization's acceptance criteria complying with CLB to assure that the level of safety will be maintained during the LTO period and providing feedback to the operating organization (if it is part of regulatory body's involvement in the LTO);

(g) Communicating with the operating organization on the progress in LTO programme activities, resolving questions and providing additional information.

In general, the regulatory bodies, along with their TSOs, have sufficient technical competence to fulfil their responsibilities with regard to the review and assessment or inspection of LTO programme(s). This includes competence to manage any contracted work (e.g. from external consultants or TSOs) effectively, and to assess the outputs produced (Requirement 20 of GSR Part 1 (Rev. 1) [1]).

8.2. DEVELOPMENT AND PROVISION OF REGULATIONS AND GUIDES

The establishment or adoption of regulations and guides to specify the principles, requirements and criteria for safety, and their promotion to interested parties, are an obligation for the regulatory body (Requirements 32, 33 and 34 of GSR Part 1 (Rev. 1) [1]). The regulations and guides are consistent with the legal system of the Member State and the nature and extent of the facilities to be regulated. The development and provision of regulations and guides are considered to be preconditions for LTO and are discussed in Sections 2 and 3.

The requirements established by the regulatory body for LTO include legal and regulatory requirements, codes, guides and international agreements and standards. They specify the requirements and associated criteria to be followed by the operating organization to ensure safety for LTO. Depending on the practice, the requirements usually refer to the scope setting methodology and presentation, ageing management, equipment qualification, maintenance, surveillance, ISI, water chemistry, AMR and, in some Member States, the organizational arrangements and knowledge management in the operating organization as well. An interface between the LTO programme and the PSR is identified, where appropriate. The requirements are developed and issued in a timely manner. The legal documents are publicly available. SSG-48 [4],

SSG-25 [7], NS-G-2.6 [20] and Ref. [8] offer elements and acceptance criteria to consider in national requirements and regulatory guides for LTO.

The operating organization may propose an alternative approach to that suggested in regulatory guidance to achieve the same safety objective. In such cases, the operating organization will typically be required to demonstrate that its proposed approach provides an equivalent level of safety. The regulatory body evaluates the acceptability of the proposals against general principles stated in laws and regulations. Usually, deviations from the regulatory guidance entail a deeper and longer regulatory review process. This approach may be used in cases where detailed requirements are not specified ahead of time.

8.3. TRAINING OF STAFF

The employment of a sufficient number of qualified and competent staff by the regulatory body is a general requirement (Requirement 18 of GSR Part 1 (Rev. 1) [1]). In general, the regulatory body staff are knowledgeable and experienced in facilitating the effective and efficient completion of the review and assessment and inspection tasks of the LTO programme. The regulatory body's management assesses the competence of the regulatory body for review of the LTO programme and, if needed, an appropriate, timely training programme is carried out for the reviewers and inspectors. Competence management and the development of training programmes are described in Safety Reports Series No. 79, Managing Regulatory Body Competence [26]. In practice, in many Member States training is delivered by the senior staff of the regulatory body or by an authorized organization. The scope and structure of the training are inserted into the general training scheme of the organization, which follows a systematic approach to training methodology [27]. The scope and duration of specific training on ageing management and LTO are in line with the competences and roles of the attendees involved in the review and assessment and inspection of LTO programmes. If necessary, the regulatory body engages TSO(s) to fill in gaps in the expertise of the regulatory staff (see GSG-12 [14]).

8.4. INTERNAL PROCESSES, PREPARATION FOR LONG TERM OPERATION PROGRAMME OVERSIGHT, REVIEW AND ASSESSMENT, AND AUTHORIZATION

It is recognized that oversight, review and assessment, inspection and authorization can be conducted in different forms of the regulatory activities (Requirements 23, 25 and 27 of GSR Part 1 (Rev. 1) [1]). Member States usually use internal processes for review and assessment and inspections related to licence renewals or the PSR process, as appropriate. The processes are described in the internal procedures within the management system. In some Member States, the regulatory body develops management projects for supporting activities related to review and assessment and inspections in the authorization process for LTO. In this case, the organization, establishment and implementation of management projects are described in the management system (Requirement 19 of GSR Part 1 (Rev. 1) [1]).

In some Member States, the regulatory body prepares and makes public a review plan or programme(s) to review and assess the LTO programme documentation or inspections to verify the compliance of the LTO programme. The regulatory plan or programme(s) contains the assessment criteria, activities to be conducted, identified submittals and their time frame, a list of the technical experts who will carry out the regulatory activities, the involvement of TSOs and other relevant information. Further information is provided in Section 8.5.

The regulatory processes and management projects can be supported by technical procedures. Where necessary, specific procedures (guides or handbooks) are developed. Good examples of these specific procedures include Refs [28–30]. Procedures for conducting the regulatory body review that describe items to be confirmed may be made public.

In general, regulatory bodies have established internal procedures for performing review and assessment, record-keeping, and documenting process performance and activities. Documents submitted for LTO by the operating organization, documents generated during the process and activity performance (review, assessment, inspection or authorization), as well as project management documents, are kept in the archive and electronic document storage database. In some Member States, the document submittals relating to LTO programmes and AMPs, outages, events and other relevant documents are stored in the licensee's databases, which the regulatory body staff can access. It is also a practice in some Member States to register the documents submitted by the operating organization or exchanged between it and the regulatory body in a shared registration database.

8.5. SPECIFIC PREPARATION OF REGULATORY BODIES REVIEWING THE PERIODIC SAFETY REVIEW FOR LONG TERM OPERATION

As mentioned in Sections 4 and 7, some Member States conduct the activities for LTO under their PSR and ageing management processes. LTO preparation is addressed as part of the PSR in the context that it concerns operation of the nuclear power plant beyond an established time frame. Thus, the PSR is a comprehensive exercise of which LTO preparation is considered a part. More emphasis is put on issues considered and work done for a plant preparing for LTO.

In this case, depending on national regulations, the regulatory body has the responsibility for different steps as described in Section 7 of SSG-48 [4], while keeping in mind a specific focus on LTO aspects, such as topics linked to safety factors 1 to 4 in SSG-25 [7]. The regulatory body is responsible for making its expectations clear in a timely manner. Regulatory oversight for PSR for LTO may go beyond what is usually performed for a regular PSR, in particular when additional activities, other than those already included in the review scope of a PSR, are required by the regulatory body for LTO, such as specific design upgrades (e.g. additional dispositions taking account of the Fukushima Daiichi accident or other major events). Further information is available in Section 3 of SSG-25 [7]. Therefore, as described in Section 8.4, the regulator's management system for PSR may need a specific update to deal with LTO.

For Member States where the PSR is the only process for justification of the safety of plant operation in the long term, it covers all the items for LTO: scope for LTO, AMR, review of AMPs and revalidation of TLAAs. Consequently, the preparation of the regulatory body (e.g. resources, competences, schedules) for LTO is to be included in the PSR preparation process. Specific attention may also be paid to the programme of corrective actions defined on the basis of the PSR in order to ensure the safety and feasibility of LTO. However, the preparatory activities for LTO for such Member States are substantially the same as for those applying the licence renewal approach.

If an integrated regulatory approach that combines PSR and licence renewal is adopted (e.g. in the practice of Armenia, Hungary and Mexico), the regulatory body needs to ensure that PSR objectives are being adequately and consistently aligned with LTO objectives.

9. OVERSIGHT OF LONG TERM OPERATION PROGRAMME PREPARATION AND IMPLEMENTATION

This section summarizes the activities and practices of the regulatory bodies for the oversight of the LTO programme of the licensees starting from preparation until the authorization of LTO.

9.1. INTRODUCTION

In Member States, regulatory bodies focus oversight activities on the preparatory and implementation phases of LTO programmes (Requirements 25 and 27 of GSR Part 1 (Rev. 1) [1], paras 3.157 and 3.220 of GSG-13 [5]). However, the timing of the oversight process, the acceptance criteria, the complexity of the review, and assessment and inspection may vary, depending on the legislative environment (e.g. licensing process for LTO, PSR without issuance of new formal licence, or a combination thereof). The goal of these activities is to ensure that the safety of the plant is not compromised during the LTO period due to ageing and obsolescence, whether anticipated at the design stage or not. The principal steps from preparation to implementation of the LTO programme and common areas of interest for regulatory bodies are described below. The subsections correspond to the recommendation on the assessment for LTO as described in SSG-48 [4].

After the requirements are established, communication with the licensee may provide further clarification. Depending on the regulatory approach, progress during the preparatory phase of LTO is overseen (e.g. organizational arrangement for the preparation of LTO programme, including the establishment of programme structure; defining the responsibilities; time schedule). At this phase, the methodology specifying the main steps of LTO preparation submitted by the licensee is typically reviewed. Regular meetings between the regulatory body and the licensee may be organized with the goal of avoiding possible misinterpretation of the regulatory requirements, identifying any difficulties and working on possible solutions together. In some Member States, preliminary results of LTO assessment by the licensee are provided to the regulatory body for comments.

After the preparatory phase, the results of LTO assessment are submitted to the regulatory body. LTO programme review by the regulatory body includes the following main steps:

(a) Checking the submitted documentation and information for its completeness against criteria given in the regulatory requirements;
(b) Conducting the review and assessment and inspections, as appropriate, against the criteria following the requirements, regulatory procedures and review plans; updating safety documentation, taking into account any public comments, where applicable;
(c) Recording review and assessment and inspection results, managing the results and communicating with the licensee;
(d) Taking a regulatory decision on continuation of operation and informing interested parties of the decision.

During review and assessment and inspection activities, communication with the licensee is maintained as appropriate, to ask, for example, for clarifications or provision of additional or missing information.Site visits are considered to be a necessary part of the review and assessment and inspection processes.An example for a standard review can be found in Ref. [28].

Depending on the regulatory framework, the submission to the regulatory body and the review of the scope setting for LTO by the regulatory body precede the submission of other documents.

Formal acceptance of the plant's preparedness for LTO may be granted in different steps of programme implementation depending on the Member State practices.

During the implementation phase of the LTO programme, primary focus is given to the following:

(a) Keeping track of conditions specified for LTO (e.g. validity of TLAAs, commitments of the licensee for the LTO period);
(b) Operating experience feedback process effectiveness;
(c) Review of organizational capability, including preparation and training of the personnel;
(d) Review of compliance with other requirements related to safety improvements for LTO, if applicable (results of PSR or other safety related reviews);
(e) Reviews and backfitting due to changes in regulatory requirements, codes and standards, state of the art;
(f) Implementation of the LTO action plan (consisting of e.g. a list of corrective actions and safety improvements, as well as a list of improvements identified by PSR, both specifying the adequate schedules), also considering the safety implications of pending actions and improvements.

9.2. REVIEW OF SCOPE SETTING FOR STRUCTURES, SYSTEMS AND COMPONENTS FOR LONG TERM OPERATION AND AGEING MANAGEMENT EVALUATION

Different requirements are used for the specification of which SSCs are included for ageing management and assessment of LTO, depending on the approach used in the Member State. For some Member States, LTO is implemented not as a specific project, but as part of 'standard' operation and oversight activities. This approach takes credit for the continuous conformity check with CLB, uses the PSR for maintaining or renewing authorization; thus scope setting for ageing management and LTO does not differ. Irrespective of the regulatory approach applied, the regulatory body provides a systematic review of scope setting according to national regulations and makes sure that the safety aspects and recommendations in SSG-48 [4] are met, so selection of the SSCs subjected to ageing management and LTO evaluation is consistent with the practice described in this reference.

If the scope of the components for ageing management and LTO is determined using the scope setting process described in Section 6.2, there may be some components that are considered to be in scope due to decisions made by the operating organization (e.g. extension of the scope to include components not directly related to nuclear safety, or other components whose failure may cause high economic loss to the nuclear power plant). These components, however, might not be covered by the regulatory review to save the resources of the regulatory body in the spirit of applying the graded approach.

A typical practice is that the licensee prepares a methodology document describing the scope setting process and submits this document to the regulatory body before or together with the result of the scope setting process. Usually, some kind of database is developed for that purpose and the regulatory body personnel is granted access to it. Both the methodology and the results are subject to regulatory review. The practice of submission or the deadline for submission of the above mentioned documents may differ depending on the regulatory approach. In some Member States, a prior agreement concerning the methodology is made between the licensee and the regulatory body.

The scope setting methodology review focuses on consistency with regulatory requirements, which in the majority of Member States are consistent with the recommendations in SSG-48 [4], as described in Section 6.2. In some Member States, a detailed review is performed for all SSCs within the scope or for a sample of SSCs (sampling is done according to predefined criteria considering a graded approach — e.g. concentrating on more important plant systems with deeper evaluation at the level of components, following risk informed principles). The regulatory experience from the oversight activities performed during previous operation is also considered. Before the review of the scope setting methodology, important information is gathered to take into account the design and other aspects of the particular plant from sources such as the following:

(a) A FSAR to identify any SSCs whose reliability and capability need to be maintained during and after the design basis event;
(b) The set of design basis events and design extension conditions;
(c) The history of failure, repair and replacement of SSCs;
(d) Other documents, such as regulatory notices or orders and licence conditions;
(e) The SSC design bases;

(f)   Other documents addressing specific areas such as fire protection, pressurized thermal shock, equipment qualification, modifications, ISI and probabilistic safety analysis;

(g)   Safety classification lists;

(h)   Operational flow charts, drawings and outlines;

(i)   Severe accident management guidelines;

(j)   Emergency plans;

(k)   Event reports;

(l)   Previous PSR results.

The availability and completeness of this necessary information, which serves as the basis for scope setting, are checked against the regulatory requirements by the review team. After gathering all important information, the scope setting methodology is reviewed, while focus is also given to justification of the methods and reference documents used to identify SSCs within the scope and to the method used to demonstrate the comprehensiveness and definition of the SSC boundaries.

The results of the scope setting process are examined to make sure that the scope setting methodology was properly applied to the master list of SSCs. The reference materials used for scope setting are also considered. Possible tools for reviewing the results of scope setting include the following:

(a)   Comparison with the list of safety class SSCs to check if all of them are within the scope;

(b)   Comparison with information provided in the FSAR (SSCs credited in safety analysis) to check if these other components important to safety are within the scope;

(c)   Review of system charts and system drawings to identify if there are other components to be included in the scope (to identify and eliminate any mistakes in the above documents to the extent reasonable);

(d)   Verification of the scope setting results by walkdowns in the nuclear power plant, where reviewers and inspectors focus on possible physical interactions of safety classified SSCs with other (also non-safety classified) SSCs located nearby and by comparing what is observable with the scope setting database (consideration of all hypothetical failures caused by interactions of the above mentioned SSCs);

(e)   Verification of the scope setting database using the results of the probabilistic safety assessment (e.g. SSCs credited in internal fire and internal flooding, external hazards, design extension conditions).

The regulatory body checks if boundaries between the in-scope and out of scope SSCs are clearly defined in the scoping methodology and applied and displayed in the appropriate lists, databases and/or drawings.

Event reports can be reviewed to identify if any SSCs concerned in the events are to be included in the scope (e.g. when a not important to safety component affected a safety related component).

In some Member States, the results of the scope setting process are required before the licence renewal application is submitted. It can be part of a submitted PSR report in other Member States.

The review makes sure that the whole scope setting process is appropriately managed, that is, that the methodology, procedures and respective documentation are controlled according to the management system of the operating organization. The regulatory body makes sure that there is an appropriate process in place to maintain the accepted and approved LTO scope for future AMRs, PSR or subsequent licence renewal.

9.3. REGULATORY REVIEW OF PLANT PROGRAMMES

The plant programmes to be reviewed by the regulatory body for LTO are those listed in Section 5.2, where detailed considerations about regulatory requirements and guides for these programmes can also be found.

The regulatory review and assessment of the plant programmes focuses on the following checks:

(a) Whether the relevant plant programmes are reviewed for consistency with the nine generic attributes of effective ageing management [4];
(b) Whether a systematic approach is used in operating the plant programmes (e.g. plan–do–check–act) [4];
(c) Whether benchmarking has been performed with other plants or international practices [8];
(d) Whether the effectiveness of the programmes is regularly (e.g. annually) assessed using, for example, performance indicators;
(e) Whether plant programmes are reviewed and updated as needed in an AMR;
(f) Whether the programmes are reviewed during a PSR (e.g. every 10 years) or if similar systematic safety assessments are conducted;
(g) Whether plant programmes are integrated under ageing management as a process in the operating organization at all stages of the lifetime of the plant (continuously);
(h) Whether the use of adequately qualified experts is ensured for analysis and assessment (continuously) in plant programmes; this includes analysis and

assessments made by TSOs, when the licensee experts are qualified to be an 'intelligent customer';

(i) Whether an effective knowledge management system is used, including knowledge transfer and know-how, within and between the respective plant programmes related to in-scope SSCs for LTO.

The plant programmes on maintenance (including repair and replacement), equipment qualification, ISI, surveillance and water chemistry need to function as interdisciplinary activities supporting the ageing management and LTO programmes and other programmes that ensure plant LTO within the safety limits. Further programme specific regulatory review and assessment aspects are provided in the following subsections.

9.3.1. Review of maintenance programme

The regulatory body reviews the scope, details and documentation of the maintenance programme. Attention is given to the following:

(a) Availability of the maintenance programmes, such as preventive and corrective maintenance for SSCs, taking into account the safety class and maintenance history;

(b) Adequacy and effectiveness of the maintenance programmes for in-scope SSCs and checking whether the licensee has performed the review of the programme for the nine generic attributes of an effective AMP;

(c) Whether the input from condition assessments, review of AMPs and revalidation of TLAAs are considered appropriately for the development or modification of maintenance programmes;

(d) Whether actual and potential ageing mechanisms are considered in the development of preventive and predictive maintenance;

(e) Whether maintenance programmes in the scope of LTO clearly identify the type of maintenance, the links with AMPs, the frequency, tasks, records and storage;

(f) Whether the programme has a systematic approach to address technical aspects such as the development of acceptance criteria, reliability centred maintenance and risk informed maintenance;

(g) Evaluation of the data collected during maintenance and their trends;

(h) Whether the documentation and database used for the maintenance related activities include preventive maintenance, condition monitoring, breakdown maintenance, replacement of components and failure history;

(i) Whether operating experience feedback is used in the maintenance programme and its outcome;

(j)   System for spare part management (as part of obsolescence), focusing on LTO;

(k)   Adequacy of technical and administrative capabilities for carrying out the maintenance activities;

(l)   Periodic evaluation of the programme to check its adequacy and effectiveness based on past experience, new knowledge and research findings;

(m)   Execution of maintenance activities by qualified personnel, after due preparation, using and following appropriate documentation, adequately maintaining the environment at the location, properly documenting the execution, taking into account already known degradation mechanisms or occurred ageing effects;

(n)   Activities that the regulatory body may be involved in, such as review of the rules and conditions to ensure the appropriateness of the programme for in-scope SSCs, approving those parts of the programme that are related to operational limits and conditions, and the changes thereto; monitoring compliance with the programme requirements, assessing the results and considering proposals for new approaches to such activities during the operational life of the nuclear power plant, including the LTO activities (see NS-G-2.6 [20]).

9.3.2.   Review of equipment qualification programme

The regulatory body reviews the scope, details and documentation of the equipment qualification programme in line with Requirement 13 of SSR-2/1 (Rev. 1) [2]. Attention is given to the following:

(a)   Adequacy of the equipment qualification programme, which includes identification of the components for qualification, including the establishment of an equipment qualification master list, identification of the environment in normal and postulated accident and design extension conditions, qualification methodology and acceptance criteria from the consideration of LTO; checking whether the results of AMR and the revalidation of TLAA have been addressed appropriately for updating the equipment qualification programme;

(b)   Whether environmental, seismic and electro-magnetic and radio frequency qualification will remain valid over the expected period of LTO or corrective measures have been developed and implemented to maintain their qualification;

(c)   Whether qualified status has been preserved and maintained through surveillance, maintenance, modifications and replacement, environmental monitoring, condition monitoring and configuration management;

(d) Whether the effectiveness of the qualification programme has been evaluated for LTO for consistency with the nine generic attributes of an effective AMP;

(e) Whether operating experience feedback has been adequately considered in the equipment qualification programme;

(f) Whether timely replacement of equipment that cannot be qualified for the planned period of LTO is adequately considered; verifying whether a specific programme for the replacement of mechanical, electrical and I&C equipment with qualified or stated lifetimes that are shorter than the planned LTO period has been developed and implemented;

(g) Availability of appropriate supply chain needed for plant modifications for LTO;

(h) Identification of additional qualification requirements, if any, by virtue of LTO and qualification of the equipment accordingly;

(i) Availability and retrievability of the equipment qualification documentation, which is to be ensured for the whole period of LTO;

(j) Whether the requalification process for equipment within the scope of LTO, which was designed in compliance with earlier standards, is focused on ensuring that the equipment can perform its function under the specified conditions;

(k) Monitoring of environmental parameters for maintaining qualified status, with special attention to identification of hotspots;

(l) System for periodic evaluation of the equipment qualification programme to check its adequacy and effectiveness.

9.3.3. Review of in-service inspection programme

The regulatory body reviews the scope, techniques and documentation of the ISI programme. Attention is given to the following:

(a) Adequacy of the SSCs included in ISI from LTO scope; verification that the results of the scope setting and AMR for LTO are adequately reflected in the existing ISI programme;

(b) Adequacy of the proposed non-destructive examinations and frequency of examination for detecting or monitoring the degradation mechanisms and ageing effects expected during LTO; checking that the evaluation provides a technical basis to justify that the ageing phenomena will be detected before they affect the safety functions of SSCs with the proposed inspection methodology and frequency;

(c) Whether internal and external operating experience and previous ISI results are appropriately considered while formulating and revising the ISI programme;

(d) Whether the plant has evaluated the effectiveness of the existing ISI programme for LTO in terms of consistency with the nine generic attributes;

(e) Availability of qualified equipment, calibration standards, procedures and qualified personnel for carrying out ISI activities, in accordance with the relevant code and national regulatory requirements;

(f) Documentation and the associated database of ISI results and their traceability in future for comparison and trending purposes;

(g) Justification of the risk informed ISI for the planned period of LTO, which needs to be verified, if used.

9.3.4. Review of surveillance programme

The regulatory review may focus on elements such as the integrity of the barriers between radioactive material and the environment, the availability and reliability of safety systems, the availability and reliability of items whose failure could adversely affect safety, and functional testing of SSCs important to safety, in line with SSG-48 [4]. Attention is given to the following:

(a) Adequacy of the surveillance programme in terms of LTO considerations (this includes SSCs covered under the surveillance programme, frequency, type of surveillance); whether the results of the AMR and scope setting for LTO are adequately reflected in the existing surveillance programme;

(b) Whether the plant has evaluated the effectiveness of the existing surveillance programme for LTO in terms of consistency with the nine generic attributes of an effective AMP;

(c) Whether operating experience feedback has been utilized for formulating and revising the surveillance programme;

(d) Whether the documentation and database of surveillance test results, the comparison and the trending of the results are available and appropriate;

(e) Whether the surveillance and monitoring programme remains effective for assessing the service life of SSCs and supporting safe LTO;

(f) Periodic evaluation of the surveillance programme for adequacy and effectiveness during LTO;

(g) Whether an additional surveillance programme for material surveillance in terms of LTO considerations is implemented (e.g. material surveillance specimens for reactor pressure vessel, pressure tubes, electrical cable samples, corrosion coupons, containment concrete samples).

9.3.5. Review of water chemistry programme

In some Member States, the regulatory body verifies that the water chemistry programme is established and implemented by the operating organization using information about the chemical and radiochemical environment to assure the integrity of structures or components within the scope of ageing management and evaluations for LTO. Attention is given to the following:

(a) Adequacy of the water chemistry programme in terms of LTO considerations; this focuses on the parameters to be monitored, frequency of monitoring, an acceptable range of parameters and an action plan based on the observed parameters. The results of AMR are to be adequately reflected in the chemistry management programme.

(b) Whether the water chemistry programme is evaluated periodically and takes input from and provides outputs to other plant programmes; for example, checking whether new findings and conclusions emanating from surveillance and ageing management are being considered in updating the plant chemistry programme and an appropriate interface is established.

(c) Whether the plant has evaluated the effectiveness of the water chemistry programme for LTO in terms of consistency with the nine generic attributes.

(d) Whether water chemistry practices are in compliance with the technical specifications, are consistent with international good practices and take account of the materials concept accordingly.

(e) Whether operating experience feedback has been utilized for establishing and revising the water chemistry programme during LTO.

(f) Checking the documentation for water chemistry parameter results and their trends.

(g) Whether infrastructure is available for the implementation of the water chemistry programme.

(h) Awareness among the chemistry staff of the implications of water chemistry parameters during LTO.

(i) Whether the chemistry programme includes control parameters that provide useful information for determining and preventing causes of unexpected ageing.

9.3.6. Review of corrective action programme

The regulatory body verifies that the corrective action programme is appropriate for the LTO period to identify and eliminate issues from safety related operating experience and avoid recurrence and that it is implemented. Attention is given to the following:

(a) Whether the effectiveness of the corrective action programmes is reviewed for LTO;
(b) Whether the corrective action programme considers ageing related degradation of in-scope SSCs;
(c) Whether the ageing aspects of external and internal operating experience are examined and appropriate corrective actions commensurate with the significance of the issues are taken, including modification of the concerned plant programmes;
(d) Whether the ageing management entity is involved in the examination of all operating experience with potential implications for ageing;
(e) Whether ageing related operating experience is documented appropriately and used as plant specific operating experience in the relevant AMP or plant programme;
(f) Whether the ageing management entity is involved in the routine review of the corrective action programme;
(g) Whether the corrective action programme is able to identify if the plant programmes credited for the ageing management of a given SSC do not adequately manage the effects of ageing and to ensure that appropriate actions, including the modification of existing plant programmes or the development of new ones, are taken;
(h) Whether the corrective action programme proactively looks for the identification of opportunities for safety improvements for the LTO period.

9.4. REGULATORY REVIEW OF AGEING MANAGEMENT PROGRAMMES AND AGEING MANAGEMENT REVIEW

The regulatory body reviews the licensee's organizational arrangements for responsibilities with respect to ageing management and LTO. The cooperation and coordination of ageing management activities, including information flow, are key considerations. The review also covers relations with external organizations, including the R&D activities of the licensee conducted in this field. An important aspect of the review of contracted activities is the question of whether the licensee has the capabilities of an intelligent customer, and whether the knowledge and results of the activities are appropriately internalized by the organization and individuals (e.g. incorporation of new knowledge into training programmes). Training programmes covering ageing management and the corresponding technical disciplines delivered to organizational units assuming responsibilities for maintaining the conditions of in-scope SSCs are also reviewed.

The requirements for and guidance on AMPs and AMR are discussed in detail in Sections 6.3 and 6.4. Regulatory review of AMR is performed within the LTO licensing process or as part of the review of the PSR results, depending on the practice in the particular Member State. Regulatory review aims to verify whether the criteria provided in regulatory documents (legislation, guides) and the licensee methodology (if it covers all important areas given by regulatory documents) have been met.

Since AMR is only conducted for in-scope SSCs, it is important for the regulatory body to ensure that the scope setting process is complete, adequate and systematic, as described in Sections 4.2 and 9.2.

It is a challenging task for the regulatory body to review the AMR performed by the licensee, since the design and the scope can include tens of thousands of SSCs in a unit. In most cases, the licensing time limits and the human resources available to the regulatory bodies necessitate a graded approach in the review of the entire AMR performed by the operating organization. A detailed review of the AMR methodology and a sampling type review of the AMR results are therefore applied to focus on specific and more important SSCs, as some Member States have reported. The regulatory body review of AMR may also require substantial technical resources; TSOs can therefore be involved in implementing this task.

When conducting the review, the regulatory body may organize technical meetings with the licensee. Some Member States have developed specific guidance or plans for the review of an AMR (even if it is part of a review within the licensing process or PSR or part of common conformity checks). The review may be carried out separately by different teams for each technical branch (mechanical, electrical and I&C, civil structures).

Specifically, regulatory review of AMR is focused on addressing the completeness of identifying the degradation mechanisms and/or ageing effects that can be relevant for in-scope SSCs operated in a particular environment and on the existence and quality of the AMPs, TLAAs or other plant programmes or analysis used for the prevention, mitigation or detection of degradation mechanisms and ageing effects to ensure the safe operation of the nuclear power plant. The regulatory body reviews the following:

(a) Whether a clear procedure for performing the AMR exists;
(b) Whether the steps for grouping are clearly described and justified in the procedure in cases where the in-scope SSCs are grouped for the AMR (commodity groups);
(c) Whether important SSC data such as design information, environment conditions, loading regimes, the results of ISI and maintenance history

(current condition of the SSCs) are accessible and were considered during the AMR and whether a system for data collection and record keeping is in place;

(d) Whether the current condition of the structures and components is assessed adequately;

(e) Whether the AMPs fulfil the nine generic attributes of effective AMPs;

(f) Whether the other plant programmes are adequate, considering the review aspects listed in Section 9.3;

(g) Whether corrective actions (modification of existing or development of new programmes) have been decided on for in-scope SSCs that are not managed, or are ineffectively managed, by an AMP, a TLAA or another plant programme, and whether the corrective actions are appropriate for the mitigation of ageing effects;

(h) Whether the AMPs are coordinated effectively with other relevant plant programmes.

The following tools and information sources can be used for the regulatory review:

(a) Design basis information and specifications;

(b) Information provided in a FSAR (e.g. information about design basis, fabrication, environmental condition and process parameters, number of load cycles);

(c) Information about past modifications;

(d) Review of updated equipment specific ageing analysis;

(e) Updated analysis of state of the art technical knowledge regarding ageing mechanisms and material degradation (including the results of R&D activities);

(f) Information received from the annual reporting of possible new degradation mechanisms in some Member States;

(g) Detailed risk informed decision making information;

(h) Comparison of AMPs, TLAAs and lists of degradation mechanisms with IGALL results [8] and other international benchmarks, including identification of missing AMPs and all important technical issues;

(i) Comparison with design specific catalogues of degradation mechanisms where they exist;

(j) Use of operating experience (sources: inspections, maintenance history, ISI results, testing and surveillance programme results, chemistry control programme results, events and component failure reports, International Reporting System for Operating Experience (IRS) and other international or national reporting systems, benchmarks);

(k) Site visits, walkdowns, inspections to obtain field evidence;

(l) Regulatory inspection findings;

(m) Implementation of complementary measures (ISI, mitigation).

Table 1 contains an assignment of the information sources and uses for an AMR.

Some of the above mentioned information is submitted to the regulatory body within the report documenting the AMR methodology and results that is submitted by the operating organization, while the others are collected by the regulatory body from its routine evaluation and inspection activities.

Focus is placed on whether the specific degradation mechanisms identified in the AMR are covered by the AMPs, TLAAs or other relevant plant programmes. Activities that need to be completed before the review may include gathering important information such as validated system drawings (validated according to real geometry, dimensions and orientation), data from manufacturer, pre-service inspections and operating data (values of certain quantities — stress, temperature, humidity). Other areas to consider are possible changes in the progress or significance of degradation due to the extended service (e.g. IGSCC, degradation caused by higher neutron fluxes).

The review of the AMPs performed by the regulatory body verifies the consistency of the AMPs with the national regulatory requirements and assures that they are appropriately coordinated and consistent with other relevant programmes.

The first step of the regulatory review (depending on review methodology) focuses on the plant level AMP, or the organization of the plant level ageing management activity if such an AMP does not exist. Another aspect is the existence and content of procedures concerning how the particular programmes are developed and improved and if the LTO phase of operation is addressed in the programme (using e.g. recommendations on the quality of AMPs and other plant programmes and on incorporating the results of AMR, as described in SSG-48 [4]; requirements on incorporating the results of AMR with respect to LTO).

The next step of regulatory review focuses on the quality and effectiveness of particular AMPs (degradation mechanism specific or component specific) and then the review focus on particular programmes (if allowed by national regulatory requirements), possibly using a graded approach or a sampling process. Comparison with with the IGALL AMPs (and the nine generic attributes specified in SSG-48 [4]) and verification of technical adequacy, consistency with the internal procedures of the operating organization concerning the development and the required content of the AMP, and management of the activities and results from AMPs and other plant programmes can be checked. For this part of the review, as well as for the review of AMR, the regulatory body uses significant

TABLE 1. ASSIGNMENT OF INFORMATION SOURCES AND USE FOR AMR

Source	Areas where used in regulatory review
Design basis information and specifications	Scope setting of SSCs SSC conditions TLAAs
Information provided in FSAR	Scope setting of SSCs SSC conditions TLAAs
Information about past modifications	Scope setting of SSCs SSC conditions
Review of updated equipment specific ageing analysis	Identification of ageing effects and degradation mechanisms Effectiveness of AMPs and other plant programmes TLAAs
Updated analysis of state of the art technical knowledge regarding ageing mechanisms and material degradation	Identification of ageing effects and degradation mechanisms AMPs and other plant programmes
Information received from the annual reporting of possible new degradation mechanisms in some Member States	Identification of ageing effects and degradation mechanisms AMPs and other plant programmes
Detailed risk informed decision making information	Scope setting of SSCs AMPs Effectiveness of other plant programmes
Comparison of AMPs, TLAAs and list of degradation mechanisms with IGALL results and other international benchmarks	Identification of ageing effects and degradation mechanisms AMPs and other plant programmes TLAAs
Comparison with design specific catalogues of degradation mechanisms where they exist	Identification of ageing effects and degradation mechanisms AMPs and other plant programmes

TABLE 1. ASSIGNMENT OF INFORMATION SOURCES AND USE FOR AMR (cont.)

Source	Areas where used in regulatory review
Use of operating experience	Scope setting of SSCs Identification of ageing effects and degradation mechanisms AMPs and other plant programmes TLAAs
Site visits, walkdowns and inspections to obtain field evidence	Scope setting of SSCs Identification of ageing effects and degradation mechanisms Effectiveness of AMPs and other plant programmes
Regulatory inspection findings	Scope setting of SSCs Identification of ageing effects and degradation mechanisms Effectiveness of corrective actions programme Effectiveness of AMPs and other plant programmes TLAAs
Implementation of complementary measures	Identification of ageing effects and degradation mechanisms Effectiveness of AMPs and plant programmes

quantities of information received from sources such as inspection and evaluation activities, and event reporting.

If a period of operation such as a prolonged outage or an extended shutdown is taking place or has taken place in the nuclear power plant, the regulatory review addresses this fact, and confirms whether specific aspects of this situation were considered in the AMR (incorporated in procedures and implemented in AMPs). In such periods, unique degradation mechanisms may occur due to the changed conditions.

9.5. REGULATORY REVIEW OF TIME LIMITED AGEING ANALYSIS REVALIDATION

The requirements for and guidance on TLAAs are discussed in detail in Section 6.5. The first challenge for regulators is to check that the TLAAs meet the definition stated in SSG-48 [4] and that the list of TLAAs is developed through a systematic approach, providing confidence that none are missing.

Revalidation of TLAAs can be a long process because it may involve an in-depth comparison with the original design analysis. The regulatory body may need to be involved in the TLAA revalidation during the development process, which could include checking the scope of analyses, reviewing and approving the general analysis methodology, inspecting the analysis process and overseeing the external organizations involved in the development. In many Member States, the regulatory body requires the operating organization to provide periodic updates on the status of TLAA revalidation.

9.5.1. Review of identification of time limited ageing analyses for definition of the long term operation programme

The first phase of the regulatory review of TLAAs usually starts with a review of the LTO programme of the operating organization. The main objective of this regulatory review is to verify the adequacy of the scope of TLAAs that will need to be revalidated by the operating organization. The regulatory body verifies whether the operating organization has performed the following activities (see also Section 6.5):

(a) Identified the list of existing TLAAs (e.g. from the FSAR and all documents that are part of the licensing basis of the nuclear power plant);
(b) Identified missing TLAAs based on the results of scope setting;
(c) Demonstrated that the new set of TLAAs is complete, for example by benchmarking with other similar plants and international good practices [8];
(d) Properly documented the existing TLAAs in the FSAR or other licensing basis documents.

In particular, in order to ensure that the TLAAs proposed by the operating organization are relevant, the regulatory body checks whether they meet all of the six criteria described in para. 5.64 of SSG-48 [4] or only the exception, as explained in para. 5.65 of SSG-48 [4].

Typical phenomena that necessitate TLAAs may include:

(a) Irradiation embrittlement of the reactor pressure vessel;
(b) Mechanical and thermal fatigue;
(c) Thermal ageing;
(d) Loss of preload;
(e) Loss of material;
(f) Change in material properties;

(g) Environmental factors necessitating qualification.

The above list is to be adapted and extended to the specific design and operation regime, while possible combinations of degradation mechanisms may need to be addressed.

9.5.2. Review of revalidation of time limited ageing analyses in the long term operation programme

The second phase of the regulatory review of TLAAs consists of a review during either the PSR or the licence renewal review phase of the LTO programme. The main objective of this regulatory review is to verify that the revalidated TLAAs confirm the maintenance of the function and safety margins necessary for the whole period of LTO and that the newly identified TLAAs are valid for the intended period of LTO. The regulatory body verifies that the TLAAs identified by the operating organization are consistent with, and meet the intentions of, the IGALL TLAAs described in Ref. [8] and cover these analyses. Based on operating experience, additional revalidation analyses can be considered.

Because the revalidation considers the degraded state at the end of the intended period of LTO, a review is carried out to verify that an evaluation has been performed for the TLAAs to demonstrate that the safety analysis meets one of the following criteria:

(a) The analysis remains valid for the intended period of LTO.
(b) The analysis has been projected to the end of the intended period of LTO.
(c) The effects of ageing on the intended function(s) of the structure or component will be adequately managed for the intended period of LTO, meaning that the operating organization can use an AMP to ensure that the TLAA remains valid during this period.

A specific focus is to review the conclusions, recommendations and suggested measures for the operating organization for LTO. Above all, in cases where a TLAA cannot be revalidated, that is, the third criterion of the above list applies, the regulatory body ensures that the operating organization has proposed appropriate corrective or compensatory actions for managing the ageing effects of SSCs during LTO, as applicable. The possible actions for this case are described in Section 6.5.

The regulatory body confirms that TLAA documentation covers the following areas:

(a) Methodology of the analyses (e.g. description of calculation model used);

(b) Criteria used for revalidation of TLAAs;

(c) Calculation of stressors and their evaluation;

(d) Whether the reviewed TLAAs justify safe operation for LTO;

(e) Calculation of residual life where appropriate;

(f) Whether the implications of revalidation are considered in the plant operational limits and conditions;

(g) Whether the TLAAs are documented in the FSAR or any other relevant reports.

The regulatory review is performed not only using the documentation provided in the LTO programme, but also through on-site inspections in order to verify that the operating organization develops and maintains all information and documentation necessary for the revalidation of TLAAs in an auditable and retrievable form. For this 'documentary' verification, the regulatory body can check the following:

(a) Accessibility of necessary design basis information, applicable codes and regulatory requirements, fabrication records, operational and maintenance history and results of inspections;

(b) Adequacy of documentation for these calculations or analyses regarding the regulator's expectations;

(c) Documentation of the revalidation of TLAAs in an update to the FSAR.

9.6. REVIEW OF TECHNOLOGICAL OBSOLESCENCE PROGRAMME

As discussed in Section 6.6, in some Member States, the regulatory body requires the licensee to establish and implement a TOP to address all SSCs important to safety and manage the spare parts required to maintain those SSCs during the design life and for the LTO of the plant. The TOP is submitted to the regulatory body for review and assessment, as applicable. The consistency of the TOP can be checked to ensure that it contains the applicable attributes set out in SSG-48 [4]. The regulatory review of the TOP may take into consideration the acceptance criteria, safety relevance, failure history, reliability of structures or components, and training to educate the personnel involved in obsolescence management. The regulatory review may involve checking that the operating organization is sufficiently proactive and that it has set up due priorities to manage obsolescence during the design life and for LTO. Attention is given to the following:

(a) Adequacy of the TOP in terms of LTO considerations; this includes checking that clearly defined roles and responsibilities relating to obsolescence programme performance and the reporting requirements for obsolescence issues are defined;

(b) Whether an overall strategic plan to mitigate some of the risks associated with electrical and I&C equipment obsolescence exists and is considered to be part of the overall lifecycle management plan;

(c) Whether the scope of the programme includes all SSCs important to safety and spare parts required to maintain those SSCs and may be applied to all SSCs important for plant reliability and availability;

(d) Whether the plant has evaluated the effectiveness of the existing TOP in terms of consistency with the nine generic attributes;

(e) Whether baseline listing of equipment and associated items in the plant exists and is managed within plant databases and information systems;

(f) Whether internal and external operating experience feedback has been utilized for formulating and revising the TOP; lessons learned and experience are shared between different plants of the same operating organization, and with other plants that have implemented a proactive obsolescence programme;

(g) Whether the TOP methodology covers the identification, prioritization and implementation of technological obsolescence solutions;

(h) Periodic evaluation of the TOP for adequacy and effectiveness during LTO.

9.7. REVIEW OF PERIODIC SAFETY REVIEW FOR LONG TERM OPERATION

The review of PSR for LTO in effect covers the review elements and aspects described in Sections 9.1 to 9.6.

A good practice for Member States considering the PSR results for LTO is to develop a dedicated oversight process for the PSR conducted by the licensee to make sure that the intent of the requirements and guides provided for the process is followed appropriately. This includes regular and ad hoc inspections, walkdowns and consultations with the licensee, as necessary. In some Member States, PSR reports (see e.g. Appendix II of SSG-25 [7]) are submitted not as a single set of documents, but in batches of documents as completed during the progress of the review. In this case, as a part of the PSR oversight process, the regulatory body may provide early comments and make requests for information to be added to the completed documents, and these can be taken into account by the licensee in the review.

Since a PSR for LTO can be more extensive, as described in Section 7, the preparations for and review by the regulatory body will also be more extensive. In most cases, this includes the review of all PSR documentation necessary for the different stages of the process:

(a) The basis document for the PSR;
(b) The safety factor report(s);
(c) The global assessment report;
(d) The final PSR report, including the integrated implementation plan.

The PSR is a major tool for identifying and, in some Member States, also deciding on the implementation of safety improvements. The regulatory body review may have a specific focus on verifying the licensee's activity, for example if appropriately broad operating experience was taken systematically into account.

The regulatory review of PSR for LTO pays special attention to safety factors 1 to 4, 8 and 9, as described in Section 7, to verify compliance with the additional requirements and guidance provided.

The regulatory body makes sure that the intended period of operation is indicated or referred to in each document meant to demonstrate the safe operation of in-scope SSCs and that the period is appropriately supported by suitable analyses or considerations.

In review of the relevant safety factors, the cumulative effects of all modifications, improvements and ageing effects, as well as obsolescence of the technology, are examined. In addition to a review of the documented results on conditions of the in-scope SSCs, regulatory walkdowns or on-scene inspection of the licensee's walkdowns can also be conducted.

A final step of the regulatory process is to review the global assessment and the corrective actions decided on with appropriate deadlines. The regulatory review may identify additional corrective actions or change the licensee's proposed ones (including the deadlines) in order to ensure the timely implementation of safety improvements.

9.8. DOCUMENTATION OF OVERSIGHT

Record keeping and documentation relating to the safety of facilities and activities are a general requirement (Requirement 35 of GSR Part 1 (Rev. 1) [1]). The management of documented information by the regulatory body follows the rules laid down in the management system. Some Member States have developed specific folders or databases to store documented information from the review and assessment process. Review and assessment performed by the regulatory body result in a decision on the acceptability of LTO.

Member States have different practices concerning the scope, structure and format of documented information from LTO review and assessment, but the common practice is that the contents of documentation from the LTO programme review and assessment are similar to the contents of other documentation from other regulatory review and assessment activities. Typically, the documentation provides the following information:

(a) Reference to the documentation submitted by the operating organization;
(b) The basis for the evaluation;
(c) The evaluation performed (evaluation records);
(d) Comparison with legal and regulatory requirements;
(e) Comparison with another similar (reference) facility or activity, where appropriate;
(f) Information on review and/or independent calculation of TLAAs performed by the regulatory body or by consultants or dedicated support organizations on its behalf;
(g) A safety evaluation report as an outcome of the review and assessment process;
(h) Regulatory findings, including evaluation of severity;
(i) Conclusions with respect to safety for LTO and continuation in operation;
(j) Additional requirements to be met by the operating organization;
(k) An oversight plan for the LTO period to confirm if compliance is maintained.

Specific technical meetings with the operating organization to discuss the results of review and assessment may be held. In such cases, a meeting summary may be prepared. Review and assessment reports are not publicly available in all Member States; in that case information may only be available to the public upon request. In general, the documentation submitted by the operating organization to demonstrate safe LTO is not publicly available in Member States.

In some Member States, the review and assessment of the LTO programme(s) are incorporated into standard periodic inspection(s) that are planned by the regulatory body; inspection(s) may also be carried out especially for LTO. The inspections are treated in the same way as other activities, and findings are documented in inspection records and/or inspection reports. The results of inspections are discussed with the operating organization, and in some Member States are submitted for licensee comments before publication. Inspection reports are publicly available in some Member States, but it is more common for them to only be available upon public request.

9.9. MANAGEMENT OF REGULATORY FINDINGS

Generic recommendations for managing regulatory findings can be found in paras 3.283–3.294 of GSG-13 [5].

In practice, in most Member States, the regulatory findings of LTO or continued operation are treated the same way as other regulatory findings. Member States use different approaches to evaluate and manage their regulatory findings:

(a) Some Member States have a risk informed decision making process or a graded approach according to safety classification or other categorization;

(b) In some other Member States, the findings are evaluated by the staff in charge of the process.

The findings are communicated to the operating organization, for example by official letters asking for additional information, or in some cases by specific technical meetings arranged with the operating organization.

In Member States that issue a new licence, the findings may be reflected in this new licence, that is, within the licence conditions.

The findings can also be used for inspections targeting ageing management or other relevant areas or programmes after authorization for LTO.

In the case of regulatory findings from inspections or from review of safety assessments, the regulatory body requests — if adequate — corrective actions as additional measures, analyses and/or documentation and sets a date for the operating organization to comply with additional requests.

If the regulatory body imposes corrective actions on the operating organization, there is usually a follow-up process to monitor progress, in particular a follow-up review after the required updates have been completed. Targeted inspections can be carried out in order to check follow-up actions such as implementation of remedial actions or fulfilment of licence conditions.

However, for Member States using limited term licences, the regulatory body usually considers that LTO related findings need to be resolved by the operating organization in a timely manner, with a target date before expiration of the operating licence.

Based on the interactions between the regulatory body and the operating organization, any remaining deviations can be resolved in different ways:

(a) The deviations can be reported to the relevant entity, such as the expert or advisory committees of the regulatory body, or the regulatory commission, where appropriate, and further licence conditions can be imposed.

(b) The operating organization can be required to implement specific actions.

(c) The decision on the licence application for LTO (where applicable) can rely on the resolution of the findings.

Where needed, the regulatory body can initiate enforcement actions for non-compliance (e.g. refer to paras 3.295–3.319 of GSG-13 [5]).

10. SPECIFIC ACTIVITIES OF THE REGULATORY BODY DURING IMPLEMENTATION OF LONG TERM OPERATION

The implementation phase for LTO follows the issuance of a formal agreement or authorization to continue operation, depending on the licensing process. Licence conditions or corrective actions from the authorization process, and commitments of the operating organization, are typically included in a plant's corrective action programme, which usually also covers other licensee activities associated with LTO. Corrective actions are short or long term in nature, depending on the character of the findings or conditions. The role of the regulatory body is to review the completeness of action plans and check the adequacy of the proposed terms for the fulfilment of corrective actions. Details on how to deal with the results are provided in Sections 9.7 and 9.8. The licensee submits the status of the action plans to the regulatory body on an agreed basis (e.g. periodically or on the achievement of milestones).

Fulfilment of corrective actions is verified by means of inspections. These inspections are carried out by the regulatory body and can be supported by external TSOs in cases of highly specialized activities or a lack of regulatory body resources. An example for post-approval inspections can be found in Ref. [30].

The scope of inspections can vary according to the items selected and can have a continuous character. Depending on the particular regulatory framework they may cover the following:

(a) Fulfilment of action plans;
(b) Ageing management, including AMPs, ageing management databases and TLAAs;
(c) Safety analyses;
(d) ISI evaluation reports;
(e) Reports on reactor pressure vessel embrittlement and brittle fracture temperature;

(f) Reports on the operational mode history of classified equipment limited by design;

(g) Maintenance reports;

(h) Timely availability of spare parts (i.e. effectiveness of TOP);

(i) Component and structure health reports.

REFERENCES

[1] INTERNATIONAL ATOMIC ENERGY AGENCY, Governmental, Legal and Regulatory Framework for Safety, IAEA Safety Standards Series No. GSR Part 1 (Rev. 1), IAEA, Vienna (2016).

[2] INTERNATIONAL ATOMIC ENERGY AGENCY, Safety of Nuclear Power Plants: Design, IAEA Safety Standards Series No. SSR-2/1 (Rev. 1), IAEA, Vienna (2016).

[3] INTERNATIONAL ATOMIC ENERGY AGENCY, Safety of Nuclear Power Plants: Commissioning and Operation, IAEA Safety Standards Series No. SSR-2/2 (Rev. 1), IAEA, Vienna (2016).

[4] INTERNATIONAL ATOMIC ENERGY AGENCY, Ageing Management and Development of a Programme for Long Term Operation of Nuclear Power Plants, IAEA Safety Standards Series No. SSG-48, IAEA, Vienna (2018).

[5] INTERNATIONAL ATOMIC ENERGY AGENCY, Functions and Processes of the Regulatory Body for Safety, IAEA Safety Standards Series No. GSG-13, IAEA, Vienna (2018).

[6] INTERNATIONAL ATOMIC ENERGY AGENCY, Communication and Consultation with Interested Parties by the Regulatory Body, IAEA Safety Standards Series No. GSG-6, IAEA, Vienna (2017).

[7] INTERNATIONAL ATOMIC ENERGY AGENCY, Periodic Safety Review for Nuclear Power Plants, IAEA Safety Standards Series No. SSG-25, IAEA, Vienna (2013).

[8] INTERNATIONAL ATOMIC ENERGY AGENCY, Ageing Management for Nuclear Power Plants: International Generic Ageing Lessons Learned (IGALL), Safety Reports Series No. 82 (Rev. 1), IAEA, Vienna (2020).

[9] INTERNATIONAL ATOMIC ENERGY AGENCY, IAEA Safety Glossary: Terminology Used in Nuclear Safety and Radiation Protection: 2018 Edition, IAEA, Vienna (2019).

[10] ORGANISATION FOR ECONOMIC CO-OPERATION AND DEVELOPMENT, NUCLEAR ENERGY AGENCY, Challenges in Long-term Operation of Nuclear Power Plants: Implications for Regulatory Bodies, OECD/NEA Nuclear Regulation NEA/CNRA/R(2012)5, OECD/NEA, Paris (2012).

[11] INTERNATIONAL ATOMIC ENERGY AGENCY, Vienna Declaration on Nuclear Safety: On Principles for the Implementation of the Objective of the Convention on Nuclear Safety to Prevent Accidents and Mitigate Radiological Consequences, INFCIRC/872, IAEA, Vienna (2015).

[12] INTERNATIONAL ATOMIC ENERGY AGENCY, The Management System for Nuclear Installations, IAEA Safety Standards Series No. GS-G-3.5, IAEA, Vienna (2009).

[13] NUCLEAR REGULATORY COMMISSION, Standard Format and Content for Applications to Renew Nuclear Power Plant Operating Licenses, USNRC Regulatory Guide 1.188, Rev. 1, NRC, Washington DC (2005).

[14] INTERNATIONAL ATOMIC ENERGY AGENCY, Organization, Management and Staffing of the Regulatory Body for Safety, IAEA Safety Standards Series No. GSG-12, IAEA, Vienna (2018).

[15] INTERNATIONAL ATOMIC ENERGY AGENCY, Licensing Process for Nuclear Installations, IAEA Safety Standards Series No. SSG-12, IAEA, Vienna (2010).

[16] INTERNATIONAL ATOMIC ENERGY AGENCY, Ageing Management and Long Term Operation of Nuclear Power Plants: Data Management, Scope Setting, Plant Programmes and Documentation, Safety Reports Series No. 106, IAEA, Vienna (2021).

[17] INTERNATIONAL ATOMIC ENERGY AGENCY, Format and Content of the Safety Analysis Report for Nuclear Power Plants, IAEA Safety Standards Series No. SSG-61, IAEA, Vienna (2004).

[18] INTERNATIONAL ATOMIC ENERGY AGENCY, Power Uprate in Nuclear Power Plants: Guidelines and Experience, IAEA Nuclear Energy Series No. NP-T-3.9, IAEA, Vienna (2011).

[19] INTERNATIONAL NUCLEAR SAFETY GROUP, Maintaining the Design Integrity of Nuclear Installations Throughout Their Operating Life, INSAG-19, IAEA, Vienna (2003).

[20] INTERNATIONAL ATOMIC ENERGY AGENCY, Maintenance, Surveillance and In-service Inspection in Nuclear Power Plants, IAEA Safety Standards Series No. NS-G-2.6, IAEA, Vienna (2002).

[21] INTERNATIONAL ATOMIC ENERGY AGENCY, Operating Experience Feedback for Nuclear Installations, IAEA Safety Standards Series No. SSG-50, IAEA, Vienna (2018).

[22] NUCLEAR REGULATORY COMMISSION, Requirements for Monitoring the Effectiveness of Maintenance at Nuclear Power Plants, 10 CFR 50.65, US Govt Printing Office, Washington, DC (2013).

[23] WESTERN EUROPEAN NUCLEAR REGULATORS ASSOCIATION, Safety Reference Levels for Existing Reactors 2020, WENRA Reactor Harmonisation Working Group (RHWG), 17 February 2021.

[24] EUROPEAN ATOMIC ENERGY COMMUNITY, Council Directive 2014/87/EURATOM of 8 July 2014 amending Directive 2009/71/Euratom Establishing a Community Framework for the Nuclear Safety of Nuclear Installations, Brussels (2014).

[25] INTERNATIONAL ATOMIC ENERGY AGENCY, Assessment of Defence in Depth for Nuclear Power Plants, Safety Reports Series No. 46, IAEA, Vienna (2005).

[26] INTERNATIONAL ATOMIC ENERGY AGENCY, Managing Regulatory Body Competence, Safety Reports Series No. 79, IAEA, Vienna (2013).

[27] INTERNATIONAL ATOMIC ENERGY AGENCY, Methodology for the Systematic Assessment of the Regulatory Competence Needs (SARCoN) for Regulatory Bodies of Nuclear Installations, IAEA-TECDOC-1757, IAEA, Vienna (2015).

[28] NUCLEAR REGULATORY COMMISSION, Standard Review Plan for Review of License Renewal Applications for Nuclear Power Plants, NUREG-1800, Rev. 2, NRC, Washington, DC (2010).

[29] NUCLEAR REGULATORY COMMISSION, NRC Inspection Manual, Inspection Procedure 71002, Licence Renewal Inspection, 2011.

[30] NUCLEAR REGULATORY COMMISSION, NRC Inspection Manual, Inspection Procedure 71003, Post-Approval Site Inspection for License Renewal, 2008.

ABBREVIATIONS

AMP	ageing management programme
AMR	ageing management review
CLB	current licensing basis
FSAR	final safety analysis report
I&C	instrumentation and control
IGALL	International Generic Ageing Lessons Learned
IRS	International Reporting System for Operating Experience
ISI	in-service inspection
ISO	International Organization for Standardisation
LTO	long term operation
PSR	periodic safety review
R&D	research and development
SALTO	safety aspects of long term operation
SSCs	structures, systems and components
TLAA	time limited ageing analysis
TOP	technological obsolescence programme
TSO	technical support organization

CONTRIBUTORS TO DRAFTING AND REVIEW

Afzal, N.	Pakistan Atomic Energy Commission, Pakistan
Ahonen, J.	Radiation and Nuclear Safety Authority, Finland
Awasthi, S.U.	Atomic Energy Regulatory Board, India
Ballesteros, A.	Joint Research Centre Petten, Netherlands
Bazso, Z.	Nuclear Regulatory Authority of the Slovak Republic, Slovakia
Bhattacarya, D.	Atomic Energy Regulatory Board, India
Billoch Colon, A.	Nuclear Regulatory Commission, United States of America
Bloom, S.	Nuclear Regulatory Commission, United States of America
Borges Araújo, J.	National Nuclear Energy Commission, Brazil
Chepurna, A.	State Nuclear Regulatory Inspectorate, Ukraine
D'Haeyer, P.	ENGIE, Belgium
Dlouha, H.	State Office for Nuclear Safety, Czech Republic
Dugan, S.	Swiss Federal Nuclear Safety Inspectorate, Switzerland
Eriksson, C.	Swedish Radiation Safety Authority, Sweden
Fors, K.	Ringhals Nuclear Power Plant, Sweden
Goicea, L.	National Commission for Nuclear Activities Control, Romania
Grey, M.	Nuclear Decommissioning Authority, United Kingdom
Grigoryan, V.	Armenian Nuclear Regulatory Authority, Armenia
Holm, T.	Oskarshamn Nuclear Power Plant, Sweden

Husarcek, J.	Nuclear Regulatory Authority of the Slovak Republic, Slovakia
Huszka, A.	Hungarian Atomic Energy Authority, Hungary
Jin, J.	Canadian Nuclear Safety Commission, Canada
Kirkhope, K.	Canadian Nuclear Safety Commission, Canada
Krivanek, R.	International Atomic Energy Agency
Makela, K.	International Atomic Energy Agency
Mansoor, K.	Pakistan Nuclear Regulatory Authority, Pakistan
Marchena, M.	International Atomic Energy Agency
Masman, F.	Forsmark Nuclear Power Plant, Sweden
Mbebe, B.	National Nuclear Regulator, South Africa
Nasarre Muro de Zaro, J.	Nuclear Safety Council, Spain
Nyisztor, D.	Hungarian Atomic Energy Authority, Hungary
Pagar, S.	Atomic Energy Regulatory Board, India
Persson, N.	Swedish Radiation Safety Authority, Sweden
Petofi, G.	International Atomic Energy Agency
Pirmoshtari, M.	Iranian Nuclear Regulatory Authority, Iran
Politi, A.	Nuclear Regulatory Authority, Argentina
Spinelli, J.	National Atomic Energy Commission, Argentina
Šupka, M.	Nuclear Regulatory Authority of the Slovak Republic, Slovakia
Synak, D.	Slovenské Elektrárne, Slovakia
Takala, M.	Forsmark Nuclear Power Plant, Sweden
Tsukabe, N.	Nuclear Regulation Authority, Japan
van Wonterghem, F.	Federal Agency for Nuclear Control, Belgium
Vaucher, R.	French Nuclear Safety Authority, France

Technical Meeting
Vienna, Austria: 1–3 October 2019

IGALL PHASE 4 Working Group 4 Meetings
Vienna, Austria: 24–27 July 2018
Budapest, Hungary: 26 February–1 March 2019
Yerevan, Armenia: 4–7 June 2019

IGALL PHASE 4 Steering Committee Meetings
Vienna, Austria: 17–19 December 2018; 9–11 December 2019

 **IAEA**
International Atomic Energy Agency

ORDERING LOCALLY

IAEA priced publications may be purchased from the sources listed below or from major local booksellers.

Orders for unpriced publications should be made directly to the IAEA. The contact details are given at the end of this list.

NORTH AMERICA

Bernan / Rowman & Littlefield
15250 NBN Way, Blue Ridge Summit, PA 17214, USA
Telephone: +1 800 462 6420 • Fax: +1 800 338 4550
Email: orders@rowman.com • Web site: www.rowman.com/bernan

REST OF WORLD

Please contact your preferred local supplier, or our lead distributor:

Eurospan Group
Gray's Inn House
127 Clerkenwell Road
London EC1R 5DB
United Kingdom

Trade orders and enquiries:
Telephone: +44 (0)176 760 4972 • Fax: +44 (0)176 760 1640
Email: eurospan@turpin-distribution.com

Individual orders:
www.eurospanbookstore.com/iaea

For further information:
Telephone: +44 (0)207 240 0856 • Fax: +44 (0)207 379 0609
Email: info@eurospangroup.com • Web site: www.eurospangroup.com

Orders for both priced and unpriced publications may be addressed directly to:
Marketing and Sales Unit
International Atomic Energy Agency
Vienna International Centre, PO Box 100, 1400 Vienna, Austria
Telephone: +43 1 2600 22529 or 22530 • Fax: +43 1 26007 22529
Email: sales.publications@iaea.org • Web site: www.iaea.org/publications

Printed and bound by CPI Group (UK) Ltd, Croydon, CR0 4YY

08/05/2026

02105777-0017